NANOTECHNOLOGY AND ENERGY

NANOTECHNOLOGY AND ENERGY

SCIENCE, PROMISES, AND LIMITS

Jochen Lambauer · Ulrich Fahl · Alfred Voß

PAN STANFORD PUBLISHING

Published by

Pan Stanford Publishing Pte. Ltd.
Penthouse Level, Suntec Tower 3
8 Temasek Boulevard
Singapore 038988

Email: editorial@panstanford.com
Web: www.panstanford.com

British Library Cataloguing-in-Publication Data
A catalogue record for this book is available from the British Library.

Nanotechnology and Energy: Science, Promises, and Limits

Copyright © 2013 Pan Stanford Publishing Pte. Ltd.

ISBN 978-981-4310-81-9 (Hardcover)
ISBN 978-981-4364-06-5 (eBook)

Printed in the USA

Contents

Notes on the Contributors

Severin Beucker is co-founder of and senior researcher at the Borderstep Institute for Innovation and Sustainability, Berlin. His research focuses on innovation and technology analyses for new technologies. In the project ColorSol, he was responsible for the analysis of market potentials and the development of application scenarios for dye solar cells.

Niels Boeing graduated in physics and science theory at Technische Universität Berlin. Since 2002 he has been working as a freelance science writer for major German publications, including *Die Zeit* and *MIT Technology Review* (German edition). In 2004 he published the popular-science introduction to nanotechnology *Nano?! Die Technik des 21. Jahrhunderts* (Rowohlt, Berlin). He lives in Hamburg, Germany.

Harald Böttner is head of the Thermoelectric and Integrated Sensor Systems department of the Fraunhofer Institute for Physical Measurement Techniques, Freiburg, Germany. He graduated with a diploma in chemistry from the University of Münster, Germany, in 1974 and received his Ph.D. in 1977 at the same university for his thesis on diffusion and solid state reaction in the quaternary semiconductor II–VI/IV–VI materials system. In 1978 he joined the Fraunhofer Institut für Silicatforschung, Würzburg, Germany, and in 1980 he changed to the present appointment at the Fraunhofer Institute for Physical Measurement Techniques, Freiburg, Germany. From 1980 to 1995 he developed IV–VI infrared semiconductor lasers, while being active in thin film thermoelectrics based on PbTe. He was one of the main inventors of the worldwide first waferscale technology for vertical thermoelectric known under "Micropelt." He is a board member of the International Thermoelectric Society, of the European Thermoelectric Society, and co-founder of the German Thermoelectric Society.

Martin Bram graduated as a materials scientist from the Friedrich Alexander University of Erlangen-Nürnberg in 1995 and received his Ph.D. from the University of Saarbrücken in 1999. After joining Forschungszentrum Jülich in 1999, the main topic of his research has been materials in energy systems. He is currently head of a research group Powder Metallurgy and Composite Materials. Dr. Bram is continuously looking for new solutions if metals or ceramics or even composites are required with defined functional porosity. His expertise in materials science and technology enables him to fruitfully combine materials synthesis, phase composition, heat treatment, grain growth, and chemical interaction during processing, as well as during operation.

Ulrich Fahl studied economics at the University of Freiburg from 1978 to 1983 and received his Ph.D. in 1990 from the University of Stuttgart on a decision support system for energy economy and energy policy. Since 1990 he heads the Energy Economics and Systems Analysis department at the Institute of Energy Economics and the Rational Use of Energy (IER) (staff of 20 researchers). He is responsible for national and international research activities in the field of energy and electricity demand and supply analysis, energy and electricity modelling, energy and environmental management in industry and commerce, sustainable development of energy systems, energy and transport issues, and energy and climate.

Regine Geerk-Hedderich is a physicist. She studied solid state physics at the Central Institute for Solid State Physics and Material Research of the TU Dresden (Dr. rer. nat.) and at the Friedrich-Schiller-Universität Jena (Dr. sc.). Between 1980 and 1989 she worked as visiting scientist for several months at the Kapitza Institute at the Academy of Science in Moscow and the High Field Magnet Laboratory in Wroclaw. During 1990–91 Dr. Geerk-Hedderich held a research position at the high field magnet laboratory in Grenoble. Since 1991 she is employed at the Karlsruhe Institute of Technology (Forschungszentrum Karlsruhe). From 1992 to 1993 she held a guest scientist position at the international superconducting laboratory in Tokyo. Since 1998 she is director the network NanoMat (supra-regional network for nanomaterials, www.nanomat.de), which has 29 partners from industry and academia.

Antje Grobe obtained her M.A. from the University of Stuttgart, Germany, where she gives lectures on dialogue management and leads several national and EU-funded research projects on risk assessment and risk perception with an emphasis on nanotechnologies and climate change issues. Grobe is managing director of DialogBasis, a science-based think-tank and dialogue platform. Since more than 15 years, she has been facilitating stakeholder dialogues and citizen participation exercises in Europe on behalf of governmental bodies, academia, industry, and civil society organizations. She serves as an expert on nanotechnologies for the European Commission and the Swiss Confederation and was with the German government's NanoKommission from 2006 to 2011.

Karl-Heinz Haas studied chemistry and obtained his Ph.D. from the University of Karlsruhe, Germany, in 1983. He joined Fraunhofer ISC (sol-gel, materials, hybrid polymers) and worked for BASF in the central polymer research lab from 1988 to 1995. In 1995 Dr. Haas became head of the hybrid polymer department at ISC. Since 2004 he is managing director of the Fraunhofer Alliance Nanotechnology and is currently also head of the New Business Development department at ISC.

Hartmut Hillmer received his Ph.D. in physics from Stuttgart University in 1989, after which he joined the Research Center German Telekom, Darmstadt. In 1991 he became a guest scientist at NTT Optoelectronic Laboratories, Japan. Since 1999 he is professor of technological electronics at the Institute of Nanostructure Technologies and Analytics, University of Kassel, Germany. In 2006 he received the Grand Prix Europeen for Innovation Award for the patent "Micro Mirror Array." Dr. Hillmer's research interests include networked sensors and actuators for smart personal environments, micromirror arrays in intelligent windows, non-invasive optical biomarker detection in breath and tissue, semiconductor lasers, and optical filters for telecommunication.

Andreas Hinsch is a physicist who has been working as a researcher for many years. He is responsible for the dye solar cell activities of Fraunhofer Institute for Solar Energy Systems (ISE), Freiburg, Germany. For the project ColorSol, he was in charge of technology research and development and the technology transfer to the companies involved.

Andreas Jäkel studied physics at the University of Kassel, Germany, from 2001 to 2008. In May 2008 he joined the Department of Technological Electronics, University of Kassel, where he worked on his Ph.D. in micro-optical and electromechanical systems with a focus on micromirror applications. He is one of the project leaders at the Institute of Nanostructure Technologies and Analytics and responsible for the development of micromirror arrays for active windows.

Jan D. König is group leader for the Thermoelectric Energy Conversion branch in the Thermoelectric Systems department of the Fraunhofer Institute for Physical Measurement Techniques (IPM), Freiburg, Germany. He is project manager in different projects regarding thermoelectric materials research, measurement systems, and thermoelectric generator development. Some of his remarkable projects include the design and fabrication of a fully automated material measurement setup, standardization of thermoelectric metrology, and the development of a small-scale production of thermoelectric generator for high-temperature application. König's current activities cover nanoscale bulk and thin-film research on Bi_2Te_3, PbTe, and silicide-based materials as well as the development of a high-temperature generator for automotive applications. Since 2009 he is executive board member of the German Thermoelectric Society.

Nico Kreinberger has a B.A. from University of Stuttgart, Germany, where he studied politics, sociology, and empirical social research. In the EU-FP 7–funded NanoCode project he conducted an international survey and several conferences on the responsible research of nanotechnologies. At the Switzerland-based Risk Dialogue Foundation he works in the fields of nanotechnologies, microsystem technologies, and climate change in several stakeholder dialogues.

Jochen Lambauer has studied environmental engineering (Dipl.-Ing., B.Sc.) at the University of Stuttgart, Germany, and the University of Iceland (Háskolí Islands, Reykjavík), Iceland. Since 2005 he is a research associate at the Institute for Energy Economics and the Rational Use of Energy (IER) at the University of Stuttgart. Lambauer is responsible for research activities in the fields of rational use of energy, energy efficiency, virtual power plants, demand response, and energy impacts of innovations (e.g., nanotechnology). In addition, he is managing director and scientific coordinator of the Graduate and Research School, Efficient Use of Energy, Stuttgart (GREES).

Claus Lang-Koetz is an environmental engineer. He obtained his doctorate degree from the University of Stuttgart. He was the manager of the group "Innovative Technologies" at the Fraunhofer Institute for Industrial Engineering IAO, Stuttgart, Germany, and coordinator of the research project ColorSol. He is now working in the machine and plant manufacturing industry as an innovation manager.

Qingdang Li studied electronics engineering at the Wuhan University of Technology, China, from 1993 to 1997, economics at the Harbin Institute of Technology, China, from 2000 to 2002, and mechanical engineering at the University of Paderborn from 2003 to 2005. In August 2006 Li joined the Department of Technological Electronics, University of Kassel, Germany, where he worked on his Ph.D. in micro-optical and electromechanical systems with a focus on micromirror applications.

Wiebke Lohstroh received her doctorate in physics in 1999 at the Georg-August Universität, Göttingen, Germany. During her stay as postdoctoral fellow at Oxford University (UK) and at Vrije Universiteit, Amsterdam (the Netherlands), she investigated structural and optical properties of thin films during hydrogen uptake. From 2005 to 2011 she worked at the Institute of Nanotechnology, Karlsruhe Institute of Technology (KIT), Germany. In 2011, she joined the Forschungsneutronenquelle Heinz Maier-Leibnitz (FRM II), TU München, Germany. Her work focuses on materials development for energy storage, i.e., solid state hydrogen storage systems and electrode materials for secondary batteries.

Wolfgang Luther works as a consultant and project manager at the VDI Technologiezentrum GmbH in Duesseldorf, Germany, since 1999. He holds a degree in chemistry, a degree in economics, and a Ph.D. in analytical chemistry. Dr. Luther's specific field of competence is the socioeconomic assessment of emerging technologies, in particular nanotechnology. His current main field of activity is the coordination of innovation accompanying measures for nanotechnology within the funding programme of the Federal Ministry of Research and Education.

Gudrun Reichenauer works in the field of materials science and physics, with a particular focus on the synthesis and characterization of aerogels and xerogels since more than 20 years. During 1999–2000 she was a research assistant in the group of Prof. G. W. Scherer at Princeton University and the Princeton Materials Institute, NJ, USA. On her return to Germany she became the head of the Nanomaterials group of the Bavarian Center for Applied Energy Research (Division: Functional Materials for Energy Technology).

Her current research is focussed on the synthesis and characterization of nanoporous materials in general, with special emphasis on sol-gel-derived materials and nanofibres synthesized by chemical vapour deposition. Application-directed activities concern, in particular, thermal insulations, electrodes in electrochemical devices, IR opacifiers, and materials for gas separation and gas storage.

Michael Steinfeldt, a diploma'd engineer, is senior scientist at the Faculty of Production Engineering, University of Bremen, Germany, since 2005. His main focus of research is environmental valuation and methods of technology assessment and life cycle assessment. Current research themes are green and sustainable nanotechnology. After some years as a process engineer in an industrial enterprise he worked as senior researcher and project manager at the Institute for Ecological Economy Research (IÖW) gGmbH, Berlin, in the field corporate environmental management (1992–2004).

Volker Viereck studied physics at the Humboldt University, Berlin, and at the University of Kassel, Germany, from 1997 to 2004. He worked out his diploma thesis at the Volkswagen Konzernforschung, Wolfsburg, on nanoparticle measurement in 2003. In June 2004 he joined the Department of Technological Electronics, University of Kassel, where he worked on his Ph.D. in micro-optical and electromechanical systems with a focus on micromirror applications. He is now leader of the Optical MEMS Technologies group there.

Alfred Voß received his Dipl.-Ing. degree in energy engineering from the Technical University of Aachen in 1970 and a Ph.D. (Dr.-Ing.) in 1973. In 1990 the University of Stuttgart appointed him director of the Institute of Energy Economics and the Rational Use of Energy. His areas of expertise are new energy technologies, including renewable energy; energy systems and energy modeling; rational use of energy; and energy and sustainability.

Wenzhong Zhu is lecturer at and manager of the Scottish Centre of Nanotechnology in Construction Materials, School of Engineering, University of the West of Scotland. His main interests and expertise are in technology and properties of self-compacting concrete and special concretes, nanotechnology in construction, and particularly nano- and micromechanical characterization of materials.

Foreword

Heat, light, and mobility are essential for our modern lifestyle. However, some of these resources, such as oil and therefore gas and diesel, are not indefinitely available. Experts on energy expect a further increase in the worldwide energy demand in the next few years. For instance, we can expect the actual numbers to double by 2050. At the same time it seems as if global oil production has already reached its maximum capacity. In order to counter the growing energy shortage, research on energy is being fostered the world over. Nanomaterials play an important role in that matter, as many of the macroscopic properties of energy materials derive from the nanoscale.

Compared with the big technological revolutions in the past, it is the small but creative ideas that nowadays spur important innovations. Knowledge gained in and through the world of nanotechnology allows us to ameliorate many existing technologies and to make them more reliable, efficient, and resource friendly. Nanotechnology will break into many different sectors, and the energy industry will see new materials with better properties come up or notice a decrease in the need of materials: high-efficiency accumulators, photovoltaics, compact fuel cells, surface coating. In the car industry we will find light-weight construction, tires with optimal adherence, self-healing varnishes, LEDs, and electro-mobility. The construction industry and process technology are two sectors that will also profit from the benefits of nanotechnology.

This book provides an interdisciplinary approach to the presentation of research results in various energy applications of nanomaterials. We look at individual technologies in their global context and deal with the resulting scientific and technological questions, commercial implementation, and ecological, ethical, and social aspects. Not only are physical-chemical basics examined, but subjects and questions concerning communication risks, protection of the environment, health, regulation or science requirements, as well as economic and social implementing are also addressed.

Storing electricity in huge quantities is one of the future challenges we will face, especially with the massive expansion of renewable energies.

To get a more precise idea of these quantities, we take a hypothetical look at the year 2030. Supposing that until then, 30% of Germany's entire electricity will be provided by wind, a storage or buffer capacity of about 3000 GWh will be necessary to make up for the energy lost during an almost wind-free week. This is more than 70 times the capacity of our actual pump storage capacity of 40 GWh. A similar problem arises in the face of a temporary energy excess. Along with pump storage plants and air pressure storages, developments in stationary storing solutions are necessary in order to store energy intelligently and to be able to feed the network when needed. Electro-chemical storage options are described in the chapter 3.3.1, "Materials for Energy Storage."

Chapter 3.4.1, "Nanotechnology in Construction," provides an overview on nanotechnology applications within the construction sector.

All over the world, scientists look for new processes in order to enhance energy and ecological assets in the cement production. CO_2 emissions in cement production are three to four times higher than, for instance, the entire air traffic's discharges. Scientists at the Karlsruhe Institute of Technology (KIT) fabricated a new adhesive agent with Celitement, which is comparable to the adhesive in Portland cement (OPC), based on the still unidentified hydraulically active calcium hydro-silicates. Compared with the standard fabrication of Portland cement, 50% of energy and CO_2 emissions can be saved during its production.

How to use lost heat efficiently with the help of the thermoelectric effect and adequate materials is the subject dealt with in chapter 3.2.2, "Nanoscale Thermoelectrics". Nanoscalic thermoelectric materials with high Seebeck coefficients show excellent characteristics for technical use — for instance, in the car industry.

Light is an elemental aspect of work quality and influences our well-being. Approximately 10% of power requirements are used for lighting appliances, half of which are employed by trade and craft businesses while 25% are used by the industry and another 25% by private households. Energy-saving lighting facilities not only aim to reduce electricity costs but set important ecological accents, a subject that is described in detail in the chapter 3.4.2, "Active Windows for Daylight Guiding Applications."

The further development of coal power plants is focused on the elimination of CO_2 by storage below the ground or below the sea level. Chapter 3.2.3, dealing with nanostructured ceramic membranes for carbon capture and storage (CCS), describes an option for technological enhancement of CO_2 elimination in power plants.

Chapter 4, on the potential analysis and assessment of the impacts of nanotechnology on the energy sector until 2030, does not only cover very interesting subjects, but completes the other chapters.

All subjects treated in this book are very important for us today, as the prevailing ecological and societal problems concern all of us. With help of new technologies and common efforts, we can create more awareness and encourage our future generations.

Dr. Regine Geerk-Hedderich
Managing Director, NanoMat

Chapter 1

Challenges in the Energy Sector and Future Role of Nanotechnology

Jochen Lambauer, Dr. Ulrich Fahl, and Prof. Dr. Alfred Voß

Institut für Energiewirtschaft und Rationelle Energieanwendung (IER), Universität Stuttgart, Germany

1.1 The Energy Sector in Germany and Its Future Challenges

An innovation in the kind of energy supply is crucial and imminent, given the finite resources and the environmental impact on climate change. Optimizing the current energy supply and energy demand patterns or developing new conventional and renewable energy technologies to meet the ever-increasing global demand for energy requires groundbreaking advances in different research areas, for example, to increase energy efficiency and adoption rate of renewable energy in the market. Novel breakthroughs in the cutting-edge field of nanotechnology, as a cross-sectional technology, show potential to be applied across the whole value chain of the energy sector (energy sources, energy conversion, energy distribution, energy storage and energy usage).

Nanotechnology and Energy: Science, Promises, and Limits
Jochen Lambauer, Ulrich Fahl, and Alfred Voß
Copyright © 2013 Pan Stanford Publishing Pte. Ltd.
ISBN 978-981-4310-81-9 (Hardcover), 978-981-4364-06-5 (eBook)
www.panstanford.com

In this context it is useful to look into the history of development of the energy demand and key parameters that influence the development of energy use and energy demand over the past years in Germany.

1.1.1 *Demographic and Economic Development*

The demographic and economic developments are important determinants of energy consumption of a country. After the reunion of Eastern and Western Germany the population grew tendentially until 2003, when it went slightly into the reverse (see Table 1.1). In 2010, the population in Germany amounted to 81.7 million. Compared with 1990, this means a population growth of around 2.3 million, or 2.9%.

The number of households grew strongly between 1990 and 2010 from 34.9 million to 40.3 million or in numerical terms around 5 million. As a result of the global financial and economic crisis a decline by 5% in GDP in Germany occurred in 2009 in comparison to 2008. In 2010 it increased again by 3.7% in comparison to 2009. Altogether economic performance grew from 1990 to 2010 by more than 29%.

Table 1.1   Demographic and economic development in Germany

	1990	1995	2000	2005	2009	2010	Change from 1990–2010
Resident population (mill.)	79.4	81.7	82.2	82.5	81.8	81.7	2.9%
Households (mill.)	34.9	36.9	38.1	39.2	40.4	40.3	15.5%
People per household	2.27	2.21	2.16	2.10	2.02	2.03	−10.9%
Living area per capita (m^2)	35.0	36.8	39.5	41.2	42.7	42.9	22.7%
Gross domestic product (bn. $€_{2005}$)	1830	1969	2159	2224	2285	2369	29.5%

Source: BMWi (2011).

1.1.2 *Development of Prices for Fossil Energy Sources*

The nineties were an era of low energy prices, and the price slumped further in consequence of the Asian financial crisis in 1997 and 1998, when the world market price for crude oil collapsed and dropped to around 12 $ per barrel. Till 2000 crude oil price rose up to 28 $/bbl. After a period of alleviation in 2004, another drastic increase occurred and reached the climax in 2008 with a crude oil price of over 130 $/bbl. The oil price increase ceased until the global banking and financial crisis in 2008, when the crude oil price fell to around 40 $/bbl till December 2008. Since then another increase in oil prices is visible, and in 2010 yearly average price for crude oil was around 77 $/bbl. This explicit increase in world prices for fossil fuels in the last years as well as the interim decrease since 2008 is reflected in the development of the consumer prices for energy and in the electricity prices in Germany.

1.1.3 *Primary Energy Consumption*

Despite growing economic performance, primary energy consumption (PEC) in Germany decreases moderately since the reunification. This is mainly due to the economic and energy-related adjustment processes in reunified Germany. Since 1995 primary energy demand was, apart from fluctuations due to temperature and economic changes, nearly constant. In 2007 PEC was around 14000 PJ (see Figure 1.1).

The obvious decrease of PEC in 2009 is mainly due to the strong increase of energy prices in 2008 as well as the following financial crisis.

PEC is still dominated by fossil fuels. Its share decreases from 87% in 1990 to 78% in 2010. Whereas the share of coal and lignite declines, the share of natural gas increases in the same time period from 15% to 22%. Demand for petroleum products in 2010, mainly driven by the traffic sector, is almost again on the same level as in 1990. The share of nuclear lowered slightly during the past years to around 1530 PJ in 2010.

Germany depends heavily on energy imports in order to cover its energy demand. The share of net imports on demand of fossil fuels increased from 53% in 1990 to 77% in 2010. The most important foreign energy supplier for Germany is Russia, with a share of almost 30% of the total energy imports.

Contribution of renewable energy to cover PEC increased from 275 PJ (1.9%) in 1995 to 1322 PJ (9.4%) in 2010 (see Figure 1.2).

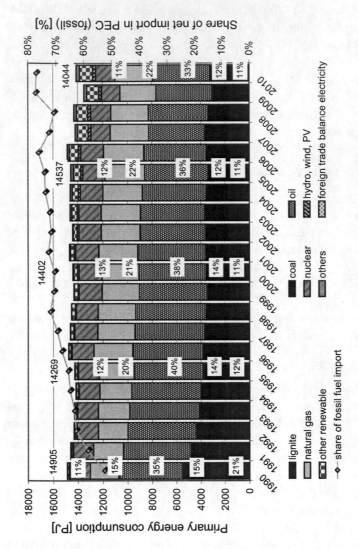

Figure 1.1 Primary energy consumpton in Germany (BMWi, 2011). See also Colour Insert.

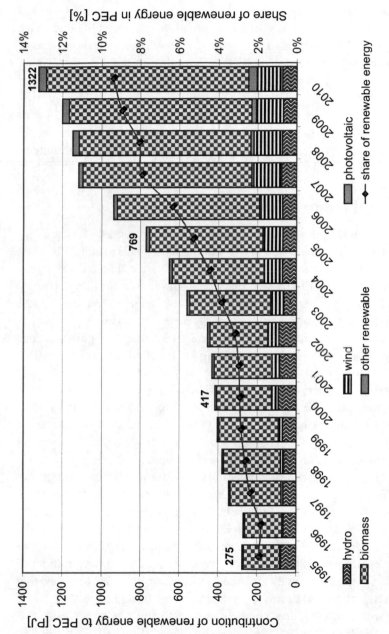

Figure 1.2 Contribution of renewable energy to PEC in Germany (BMWi, 2011). See also Colour Insert.

A considerable amount of this increase is attributable to biomass that is mainly used for heat generation and evermore for the production of bio fuels. Biomass, together with the biogenic share of waste, amounts to over 78% of the energy generation based on renewable energy. Another 10% applies to wind energy, which shows a high rate of growth starting in 2000. Use of solar energy in the form of solar heat and photovoltaics clearly increased during the past years. Its contribution to cover energy demand is still very marginal. The same applies to energy generation by geothermal energy. Energetic use of hydro power was around 77 PJ already in 1995 and could not be increased since then.

1.1.4 *Electricity Generation*

Since 1995 gross electricity consumption increased after a slight growth in the beginning of the nineties, and it was around 610 TWh in 2010 and therefore 69 billion kWh higher than in 1995 (see Figure 1.3). Similarly, electricity generation was declining strongly in 2009. This was due to the lower energy demand in industry and the tertiary sector resulting from the economic crisis.

With a share of 23.2%, 22.4% and 18.6% in 2010 lignite, nuclear and coal were the main supporting pillars of electricity generation. In chronological sequence since 1990 their share on the total electricity generation declined slightly. In comparison the share of natural gas doubled from 6.5% in 1990 to 13.8% in 2010. In 2010 57% of electricity generation was based on fossil fuels compared with 65% in 1990. The share of electricity generation based on combined heat and power is 16% at present.

Through the funding constituted by the Renewable Energy Act (EEG), the gross electricity generation with renewable energy grew since 2000 considerably and quadrupled nearly from 29 billion kWh in 1995 to 109 billion kWh in 2010 (see Figure 1.4).

Whereas economic extension of hydro power in Germany is more or less impoverished, electricity generation by wind power and biomass could be strongly extended. In 2010 around 35% of renewable electricity generation was by wind power. In regard to the rate of growth photovoltaic showed the highest values between 1995 and 2010, although solar-based electricity generation had a share of just 11% of the total renewable electricity generation in 2010.

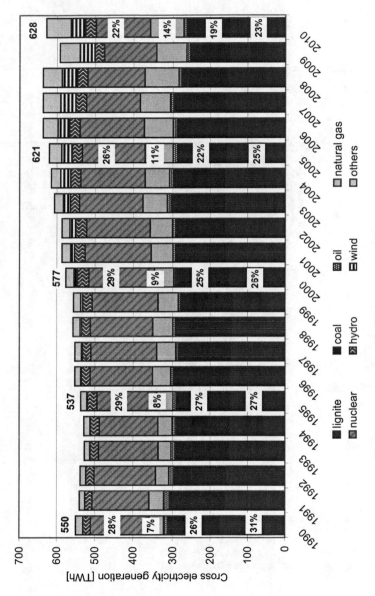

Figure 1.3 Cross electricity generation in Germany (BMWi, 2011). See also Colour Insert.

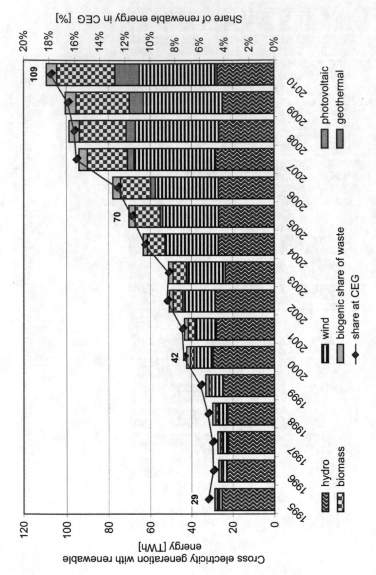

Figure 1.4 Contribution of renewable energy to the cross electricity generation in Germany (BMWi, 2011). See also Colour Insert.

Altogether the share of renewable energy on gross electricity generation was around 17% in 2010.

German power plant fleet included an electricity generation capacity of around 170 GWel at the end of 2010, with the following share of different energy carriers as well as technologies: coal 18%, wind energy 16%, natural gas 14%, lignite 13%, nuclear 13%, hydro power 6%, petroleum products 3% and others 17%.

1.1.5 *Final Energy Consumption*

Since reunification final energy consumption in Germany showed a slight decrease from 9472 PJ in 1990 to 9242 PJ in 2005 (see Figure 1.5). The share of the different sectors on the final energy consumption showed just little changes over the past years. The share for the industrial, households and transport sectors in 2010 was 28%. The share of the tertiary sector decreased since the nineties tendentially and was around 15% in 2010.

Final energy consumption is still dominated by petroleum products, mainly in the form of fuels. The contribution of renewable energy to the final energy consumption in 2010 was about 11%. This comprises the direct use of renewable energy as well as the use in the form of regenerative electricity, bio fuels and district heating based on renewable energy sources.

1.1.6 *Energy Productivity and Energy Intensity*

Improvements in energy efficiency are one of the most important issues in the ongoing discussion in terms of energy policy. Table 1.2 shows an overview of important parameters for the development of energy efficiency in energy conversion as well as in the different final demand sectors.

For the energy conversion sector, the parameter average utilization factor of fossil fuel-fired electricity generation plants is used. This parameter increased from 1991 to 2010 by around 22%. In the nineties the replacement of inefficient generation plants in Eastern Germany was one main reason for this increase.

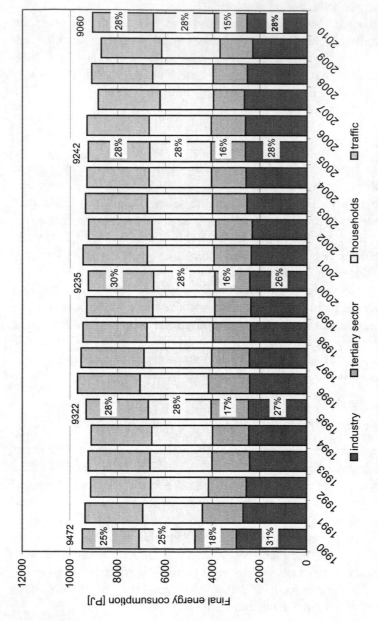

Figure 1.5 Final energy consumption in Germany (BMWi, 2011). See also Colour Insert.

Table 1.2   Parameters for energy intensity in different sectors

		1991	1995	2000	2005	2009	2010	Relative change 1991–2010
Generation and conversion	average utilization factor of fossil fuel-fired electricity generation plants (%)	37.4	38.1	39.7	42.2	44.6	45.6	21.8%
Industry	FEC (TJ)/bn. €$_{2005}$ gross domestic product	1438.0	1256.5	1121.4	1173.5	1009.3	1072.9	−25.4%
Tertiary sector	FEC (TJ)/thousand employee	65.2	56.8	49.6	48.2	44.3	43.9	−32.7%
Households	FEC (kWh)/m^2 living area	242.8	245.5	249.7	223.9	211.8	193.9	−20.1%
Traffic	fuel consumption/vehicle and 100 km mileage	9.2	8.8	8.3	7.8	7.5	–	−18.5%*

*Relative change 1991–2009
Source: (BMWi 2011).

In industry final energy consumption per unit GDP was reduced by around 25% in the time period from 1991 to 2010. The main reason was a considerable reduction in specific fuel usage, whereas in the same time specific electricity consumption in industry showed only a slight increase as a result of the growing share of electricity in the final energy consumption. Furthermore, structural changes in industry towards less energy-intensive branches of industry play an important role in the enhancement of energy productivity.

In households over 70% of the final energy consumption is in the form of space heating. Therefore, final energy consumption per square meter living area is often used as parameter to evaluate energy efficiency for this sector. Based on significant improvements in insulation as well as by the use of more efficient heating systems, specific energy consumption per unit living area could be reduced from 1991 to 2010 by around 20%.

Final energy consumption of the tertiary sector is comparatively heterogeneous in which the number of employees is a major parameter in this sector. The reduction of final energy consumption per employee could be reduced by 33% from 1991 to 2010.

One of the determining factors was similar to the household sector, a considerable reduction of the specific space heat demand and on the other hand similar to the industrial sector, a growing share of electricity on the final energy consumption.

The average fuel consumption per passenger car and 100 km mileage is used as indicator for the development of energy efficiency in the traffic sector. This value shows a decrease of 18% in the period from 1991 to 2009. The reason for that are essential technical improvements in new vehicles as well as a strong increase in the number of diesel vehicles. In comparison with petrol-driven vehicles they show notable lower specific fuel consumption. Another reason is the trend of fuel-efficient compact vehicles.

1.1.7 *Emissions*

In the nineties emissions of important air pollutants could be reduced drastically. Between 1990 and 2009 energy-related SO_2 emissions could be reduced by around 93%. In the same time period NO_x, CO and PM could be reduced by 58%, 77% and 93%, respectively. The main reasons were flue gas cleaning in power plants, the reduction of sulfur content in fuels and the area-wide introduction of catalyst for vehicles.

At the moment the focal point of environmental politics lies in the reduction of greenhouse gas (GHG) emissions. Around 81% of current GHG emissions are due to extraction and generation of energy. Between 1990 and 2010 the energy-related CO_2 emissions could be reduced by around 20% from 977 million tons to 777 million tons (see Figure 1.6).

Between 1999 and 2004 energy-related CO_2 emissions varied around 800 million tons. Since 2005 there was another slight decrease visible, supposedly due to high energy prices and the introduction of emission trading. With around 45% almost half of the energy-related CO_2 emissions in 2010 resulted from the energy sector (generation and conversion), 20% from households and the tertiary sector, another 20% from the traffic sector and around 15% from emissions in industry.

With a decrease of almost 36% from 1990 to 2010 the industry sector showed the highest reductions. CO_2 emissions of energy generation and conversion could be reduced in the same time period by around 19%. In the traffic sector a reduction of only 5% was possible. Energy-related CO_2 emissions per inhabitant decreased in the same time period by around 23% and was around 9.5 tons per annum in 2010. CO_2 output per GDP could be reduced by 39% between 1990 and 2010.

It can be concluded that the world has been increasingly powered by fossil fuels over the last 150 years. Surging oil and gas prices have drawn physical and political attention to curbing rising production and to substantiating the importance of affordable supplies to the world economy. The energy security concerns are compounded by the increasingly urgent need to mitigate GHG emissions – the bulk of which come from burning fossil fuels – to prevent non-tolerable damage to the global climate. There is growing consensus that continuing along the current path of energy system development is not compatible with sustainable development objectives. It is a critical challenge to find ways to expand the quality and quantity of affordable energy services for a growing population while simultaneously to effect a rapid transformation to a low-carbon

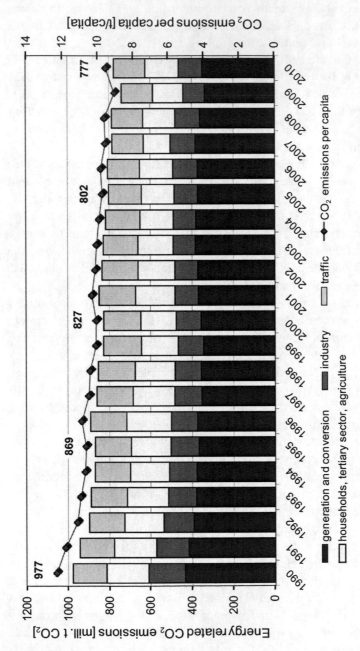

Figure 1.6 Energy related CO_2 emissions in Germany (BMWi, 2011). See also Colour Insert.

and environmentally benign system of energy supply. Some energy technologies include the following:

- harnessing alternative sources of energy,
- decarbonizing power generation and
- improving the energy efficiency in buildings, appliances, transport and industry.

They are the keys for a transition to a sustainable energy system in the future.

The scope of this book is to describe, analyse and evaluate how nanotechnology could meet these future challenges. It gives an overview of nanotechnological applications within the value chain of the energy sector and evaluates selected nanotechnological applications which could have significant impacts on the energy sector directly and indirectly.

Furthermore, the book gives a comprehensive description of the impacts and outcomes of chosen nanotechnological applications on energy consumption, energy sources, energy supply and the energy industry in Germany and shows the potential of these applications for energy savings, improvement in energy efficiency and the reduction of emissions until 2030.

1.2 Nanotechnology and Energy

Nanotechnology can be applied to the whole value chain of the energy sector: energy sources – energy conversion – energy distribution – energy storage – energy use. An overview of the possible application range of nanotechnology in the energy sector (see Figure 1.7) shows the vast and broad variety of possible products and applications. This variety can be particularly exemplified by Aerogels, which can be applied to supercapacitors, used as functional component in fuel cells, as catalyst support or as thermal insulation in almost every temperature regime. Nanotechnology has the potential for energy efficiency in various branches and industries by applying new technological solutions and product technology optimization. Nanotechnology can also benefit renewable energy sources and facilitates an economical

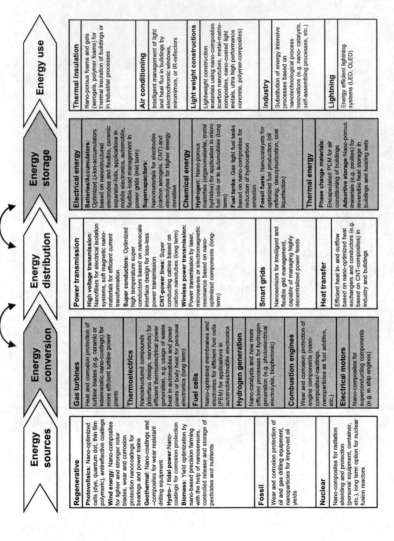

Figure 1.7 Possible applications of nanotechnology in the energy sector (Luther, 2008).

breakthrough. Altogether, nanotechnology can make essential contributions to a sustainable energy system and to a worldwide climate protection policy in the long run.

Nanotechnology offers essential improvement potentials for the exploitation of conventional energy sources (fossil, nuclear) and the use of renewable energy sources (solar, wind, geothermal, hydro, tidal or biomass). For example, nano-coated wear-resistant drill bits optimize lifetime and efficiency of oil and gas production or geothermal energy facilities. Other examples are high-performance nanomaterials for the use in wind turbines and tidal power plants as well as corrosion- and wear-resistant coatings for mechanically stressed components (bearings, transmissions). In the photovoltaic sector, nanotechnology can play an essential role. With the help of anti-reflective coatings, it is possible to enhance the energy efficiency of conventional crystalline silicon solar cells. Most notable is the further development of alternative technologies such as thin film solar cells, dye solar cells or polymer solar cells. The main focus lies in the increase in efficiency of alternative photovoltaic technologies. In the long term, the utilization of quantum dots and quantum rods could achieve an efficiency of more than 60%.

In the field of energy conversion nanotechnology could lead to great reductions in CO_2 emissions by efficiency enhancements at conventional power plants. A precondition is an increase in efficiency and therewith working temperature. Therefore the development of heat-resistant turbine materials is necessary. Improvements can be achieved by heat and corrosion protection coatings on the nanoscale that allow higher working temperatures and the use of light-weight materials (e.g., titanium aluminide). In the field of carbon capture and storage (CCS), nano-optimized membranes can offer the possibility of capturing and storing CO_2 with lower efficiency losses. Another promising possibility of generating energy is thermoelectric generators. By optimizing the interface design of nanostructured semiconductor and efficiency enhancements, wide application in power plants, industry or automotive industry is possible.

The application of nanomaterials (e.g., carbon nanotubes) has the prospect of reducing energy losses during power transmission. Therefore the extraordinary electric conductivity of nanomaterials should be employed. Furthermore there are nanotechnological approaches to optimizing supraconducting materials and consequently enabling a loss-free power transmission. Nanotechnology could contribute to future grids that allow dynamic load, failure management, demand-driven energy supply with flexible price mechanism as well as support of local renewable energy sources supply. For example, nanosensors and power electronic components could be used to control and monitor smart grids in the future.

Nanotechnology has so far been successfully employed to improve electric energy storage devices (accumulators, batteries) and was able to improve the security and performance of Li-ion accumulators by using innovative, ceramic, heat-resistant but still flexible separators and high-performance electrode material. In the long term, it is expected that improvements in storing hydrogen can be viewed as a promising energy source for environmentally friendly energy supply. Current materials do not meet the requirements and standards that are, for example, asked for from the automotive industry. Nanomaterials, for example, nanoporous metal organic compounds, offer vast potential for new developments in this field. Another area of energy storage devices are thermal energy storages. These include phase change materials (PCM) for latent heat storage systems to reduce space heating and cooling demand of buildings. While adsorption storage systems based on nanoporous materials are economically attractive. For instance, zeolites employed as heat storage systems in industry or in district heating.

Furthermore nanotechnology offers multiple possibilities for reducing energy consumption. For example, in the transport sector, light-weight materials based on nanocomposites, optimized fuel combustion with the help of additives, wear-resistant coatings, lighter engine components or nanoparticles in tires can reduce energy demand. In industrial processes, tribological coatings for mechanical components in equipment and machinery can result in energy savings, while in the field of construction, there is a great potential for energy savings by employing nanoporous insulation materials. In general the guiding of heat and light fluxes by nanotechnological components (micromirrors, switchable glazing windows) is a promising field to reducing energy demand of buildings in new constructions as well as in existing buildings (Luther,

2008). For references and literature about possible applications of nanotechnology, please refer to the works by Greiner *et al.* (2009); Haas and Heubach (2007); Heubach *et al.* (2005); Luther (2007a); Luther *et al.* (2011); Noack (2007); Pastewski *et al.* (2009); Wagner and Zweck (2006); Werner *et al.* (2006) and in particular Luther (2008) for the energy sector.

Bibliography

BMWI (2011). "Zahlen und Fakten: Energiedaten – Nationale und Internationale Entwicklung, Stand: 07.12.2011." Bundesministerium für Wirtschaft und Technologie.

Greiner, F., Schlaak, H. F., Tschulena, F., and Korb, W. (2009). Mikro-Nano-Integration – Einsatz von Nanotechnologie in der Mikrosystemtechnik, HA Hessen Agentur GmbH, Wiesbaden.

Haas, K.-H., and Heubach, D. (2007). NanoProduktion – Innovationspotenziale für hessische Unternehmen durch Nanotechnologien in Produktionsprozessen, HA Hessen Agentur GmbH, Wiesbaden.

Heubach, D., Beucker, S., and Lang-Koetz, C. (2005). Einsatz von Nanotechnologie in der hessischen Umwelttechnologie: Innovationspotenziale für Unternehmen, HA Hessen Agentur GmbH, Wiesbaden.

Luther, W., Beismann, H., Seitz, H. (2011). Nanotechnologie in der Natur - Bionik im Betrieb, Zukünftige Technologien Consulting, VDI Technologiezentrum GmbH, Hessen Agentur GmbH, Wiesbaden.

Luther, W. (2007a). Einsatz von Nanotechnologien in Architektur und Bauwesen, HA Hessen Agentur GmbH, Wiesbaden.

Luther, W. (2008). Einsatz von Nanotechnologien im Energiesektor, HA Hessen Agentur GmbH, Wiesbaden.

Noack, A. (2007). Nanotechnologien für die optische Industrie – Grundlage für zukünftige Innovationen in Hessen, HA Hessen Agentur GmbH, Wiesbaden.

Pastewski, N., Lang-Koetz, C., Heubach, D., and Haas, K.-H. (2009). Materialeffizienz durch den Einsatz von Nanotechnologien und neuen Materialien, HA Hessen Agentur GmbH, Wiesbaden.

Wagner, V., and Zweck, A. (2006). Nanomedizin – Innovationspotenziale in Hessen für Medizintechnik und Pharmazeutische Industrie, HA Hessen Agentur GmbH, Wiesbaden.

Werner, M., Kohly, W., Simic, M., Forchert, C.-E., Rumsch, W.-C., Klimpel, V., and Ditfe, J. (2006). Nanotechnologien im Automobil – Innovationspotenziale in Hessen für die Automobil- und Zuliefer-Industrie, HA Hessen Agentur GmbH, Wiesbaden.

Chapter 2

Principles of Nanotechnology

Nanotechnology is regarded as a key technology for innovation and technological progress in almost all branches of industry and economy worldwide. Novel breakthroughs in the cutting-edge field of nanotechnology, as a cross-sectional technology, show potential to be applied to the whole value chain of the energy sector. This chapter provides the definition of nanotechnology and describes nanomaterials, production processes, and tools of nanotechnology.

2.1 Definition and Classification

Jochen Lambauer, Dr. Ulrich Fahl, and
Prof. Dr. Alfred Voß
*Institut für Energiewirtschaft
und Rationelle Energieanwendung (IER),
Universität Stuttgart, Germany*

Nanotechnology cannot be clearly and unanimously defined, owing to its scientific basics that can be found in the convergence of different natural sciences, such as physics, chemistry, and biology. Furthermore, knowledge of engineering science, for example, mechanical engineering or electrical engineering, is used (Hullmann,

Nanotechnology and Energy: Science, Promises, and Limits
Jochen Lambauer, Ulrich Fahl, and Alfred Voß
Copyright © 2013 Pan Stanford Publishing Pte. Ltd.
ISBN 978-981-4310-81-9 (Hardcover), 978-981-4364-06-5 (eBook)
www.panstanford.com

2001). Nanotechnology is an umbrella term for a variety of technologies dealing with structures and processes on the nanometre scale. One nanometre is a billionth of a metre (10^{-9} m) and indicates a threshold where properties of matter can no longer be described with the principles of classical physics but where quantum effects play a more and more integral role (Paschen *et al.*, 2004). As early as in 1959, an American Nobel Prize winner in physics, Richard P. Feynman, referred to the systematic manipulation of matter on the atomic scale as a guiding vision of nanotechnology.

'But I am not afraid to consider the final question as to whether, ultimately – in the great future – we can arrange the atoms one by one the way we want them. [...] What would the properties of materials be if we could really arrange the atoms the way we want them? [...] Atoms on a small scale behave like nothing on a large scale, for they satisfy the laws of quantum mechanics. [...] At the atomic level, we have new kinds of forces and new kinds of possibilities, new kinds of effects. The problems of manufacture and reproduction of materials will be quite different. [...] The principles of physics, as far as I can see, do not speak against the possibility of maneuvering things atom by atom. It is not an attempt to violate any laws; it is something, in principle, that can be done; but in practice, it has not been done because we are too big' (Feynman, 1960).

There exists up to now no generally accepted definition of nanotechnology. This is because nanotechnology is a heterogeneous field of technology on the one hand, and on the other hand, the distinction to other adjacent areas such as microelectronics, chemistry, and biology aggravates a generally admitted definition. Within the scope of this book the definition from the German Federal Ministry of Education and Research is used: 'Nanotechnology describes the creation, analysis and application of structures, molecular materials, inner interfaces and surfaces with at least one critical dimension or with manufacturing tolerance (typically) below 100 nanometers. The decisive factor is that new functionalities and properties resulting from nanoscalability of system components are used for improvement of existing products or the development of new products and application options. Such new effects and possibilities are predominantly based on the ratio of surface-to-volume atoms and on the quantum-mechanical behaviour of the elements of the material' (Schulenburg, 2006). The definition of nanotechnology can be divided into two criteria. The first criterion characterises the geometrical scale and describes a necessary condition. The limitation of 100 nm (subject to variations between literatures) is used in the context of this book. According to Köhler, (2001), a nanostructure in a strict sense must show two dimensions smaller than 1 nm. For nanostructures in a wider sense, he insists that one dimension is smaller than 1 nm and the second dimension is smaller than 1 µm. The definition of the extent of the second dimension is of relevance because it decides if micro-structured ultrathin layers with just one dimension in nanoscale count as nanotechnology. The second criterion specifies that not just the geometry is a determinator, but above all the activation of new effects and properties is essential for nanotechnology (Paschen *et al.*, 2004). Despite these two criteria there is a variety of border cases that allow elbowroom for definition and they impede a strict distinction. In its recommendations from October 18th 2011 the European Commission defined the term 'nanomaterial' as follows: 'Nanomaterial means a natural, incidental or manufactured material containing particles, in an unbound state or as an aggregate or as an agglomerate and where, for 50% or more of the particles in the number size distribution, one or more external dimensions is in the size range 1 nm −100 nm. In specific cases and where warranted by concerns for the environment, health, safety or competitiveness the number size distribution threshold of 50% may be replaced by a threshold between 1 and 50%. By derogation from the above, fullerenes, graphene flakes and single wall carbon nanotubes with one or more external dimensions below 1 nm should be considered as nanomaterials' (EU, 2011).

Nanotechnology attracts worldwide more and more attention and is indicated to be a future technology (Schulenburg, 2006). Pervasive innovations are expected from nanotechnology in almost all technological fields (Paschen *et al.*, 2004). Richard E. Smalley, co-discoverer of fullerenes and Nobel Prize winner in chemistry, characterised the impacts of nanotechnology on health, prosperity, and living standard as follows:

'The impact of nanotechnology on health, wealth, and the standard of living for people will be at least the equivalent of the combined influences of Microelectronics, medical imaging, computer-aided engineering, and man-made polymers in this century' (see Paschen *et al.*, 2004). Even though a lot of applications of nanotechnology are in the early developments, it can already be stated that it is no hype but the technology for the 21st century (Boeing, 2005). This statement can be best supported by two successful examples, light-emitting diodes (LED) and platinum nanoparticles. The former is based on nanometre-thick semiconducting layers and has captured already a high-volume mass market (Luther *et al.*, 2004). The latter can be used as catalysts in fuel cells in power engineering and have been well established (Steele and Heinzel, 2001). Compared to the promising possibilities and potentials of nanotechnology, there is still insufficient data concerning the possible toxic impacts of nanoparticles on organisms available. Nevertheless, it is inappropriate to stigmatise and characterise nanotechnology wholesale as hazardous, and neither is it possible nor desirable to decelerate or prevent the utilisation of nanotechnology, because the potentials for a sustainable energy system, for a sparing use of resources or for healing of deadly diseases, are, according to (Boeing, 2005), enormous.

Bibliography

Boeing, N. (2005). Nanotechnik, *Tech. Rev.*, 32–41.

EU (2011). Commission Recommendation of 18 October 2011 on the definition of nanomaterial, Europäische Kommission, Official Journal of the European Union, (2011/696/EU), Brüssel.

Feynman R. P. (1960). There's Plenty of Room at the Bottom An Invitation to Enter a New Field of Physics, *Caltech Engineering and Science*, Nr. 23:5, S. 22–36.

Hullmann, A. (2001). Internationaler Wissenstransfer und technischer Wandel - Bedeutung, Einflussfaktoren und Ausblick auf technologische Implifikationen am Beispiel der Nanotechnologie in Deutschland, Physika-Verlag, Heidelberg.

Köhler, M. (2001). Nanotechnologie: Eine Einführung in die Nanostrukturtechnik, Weinheim u.a.

Paschen, H., Coenen, C., Fleischer, T., Grünwald, R., Oertel, D., and Revermann, C. (2004). Nanotechnologie - Forschung, Entwicklung, Anwendung, Springer Verlag, Berlin, Heidelberg, New York.

Schulenburg, M. (2006). Nanotechnologie: Innovationen für die Welt von Morgen, Bundesministerium für Bildung und Forschung (BMBF), Bonn, Berlin.

Steele, B. C. H. and Heinzel, A. (2001). Materials for fuel-cell technologies, *Nature*, 414(6861), 345–352.

2.2 Scientific and Technical Background

Jochen Lambauer, Dr. Ulrich Fahl, and
Prof. Dr. Alfred Voß
*Institut für Energiewirtschaft
und Rationelle Energieanwendung (IER),
Universität Stuttgart, Germany*

In the following sections, different types of nanomaterials as well as a selection of tools and production processes are described in detail. Furthermore two different strategies to reach 'nanolevel' are examined.

2.2.1 *Nanomaterials*

Nanotechnology is constituted by geometrical structures of atoms or molecules. Thereby it is distinguished between point-shaped structures, line-shaped structures, layer structures, pores structures, and complex structures. Table 2.1 shows examples of nanomaterials and their application spectrum.

2.2.1.1 Point-shaped structures

Point-shaped structures with all three dimensions in nanoscale are classified as crystalline or amorphous nanoparticles, nano-islands, bigger molecules, or clusters. Their properties can differ apparently from solid bodies made from the same material. For instance, metals can become isolators if their particle size is very small. It is characteristic for point-shaped structures that they show properties which are located in a range between the properties of solid bodies and molecules. Nanoparticles have a relatively huge surface in comparison to their volume. In the case of a nanoparticle with a diameter of 10 nm, around 20% of the atoms are located on its surface. It can increase to more than 90% when the nanoparticle

Table 2.1 Examples of nanomaterials and their application spectrum (Haas and Heubach, 2007)

Form	Materials	Application examples
Particle (filler)	TiO_2, SiO_2, Al oxide, ZnO, stibium tin oxide, indium tin oxide, C_{60} fullerene, Ag	Fillers for composites, e.g., nanoparticles for lacquer, transparent electrical conductivity, antimicrobial functions
Moulded bodies	Metals, ceramics, nanoparticles in metallic or ceramic hard layers	Ceramic or metallic moulded bodies with increased hardness or viscosity
Composite	Nanoparticle in polymers	CNTs for mechanical reinforcement, layer silicates in polymers for improved barrier properties (flame protection)
	Nanopores	Heat insulation, anti-reflective layers
Fabrics/tubes	CNTs, polymer nanofabrics	Fillers for composites (improved electrical conductivity, heat conductivity), filtration
Layers	Nanolacquer, anorganic-organic hybridpolymers, single or multiple layers	Scratch-resistant layers for polymers, tribological protection layers for metals, photocatalytic active layers, anti-microbial layers
	Structured surfaces (etching structures, embossing)	Optics, anti-reflective layers, diffractive optical elements, lotus-effect structures for reduced surface energy

Abbreviation: CNT, carbon nanotube.

has a diameter of 1 nm. Optical and electronic properties of nanoparticles depend mainly on the size of the particles. It is, for example, possible to adjust the fluorescence colour of a cadmium telluride particle from green to yellow and red by changing the size. Such nanocrystals are suitable as markers in biological systems or can be used for manufacturing LEDs (Paschen *et al.*, 2004).

As a rule, it is not necessary for nanoparticle atoms to be in a definite number, and therefore they often show dispersion in particle size. In comparison, clusters have a defined number of atoms (Köhler, 2001). A cluster stands for an organised aggregation of atoms containing one or several chemical elements for a unit of solid particle. Unlike metallic or semiconductive crystals, clusters show, similar to atoms or molecules, discrete electric conditions. For this reason, clusters are sometimes misleadingly called nano-atoms or quantum dots (Grüne *et al.*, 2005).

Such quantum dots are nanoparticles based on semiconductor material. If the size of the semiconductor materials is small enough, quantum effects appear that limit the energy levels, and thereby electron activities and electron holes can exist. The optical properties of quantum dots are adjustable simply by altering the size, because energy bears a relation to wavelength (or colour). Therefore, particles can be manufactured that radiate or absorb light at a particular wavelength (colour). Such kind of quantum dots can be used, for example, in dye solar cells or in quantum dot LEDs (RS, 2004).

An important material class with superior properties and special structures represents fullerenes. Fullerenes are made up of five- and six-cornered, more or less round carbon cage molecules in different sizes. They have small densities (ca. 1.5 g/cm^3) and are electric non-conductors (Binnenwies *et al.*, 2003). They offer the possibility to encapsulate external atoms in their interior. Some of their superior properties include almost flawless rotational symmetry and high mechanical stability, and they offer the possibility to bind functional groups on the molecule at spatially selected and defined locations (Paschen *et al.*, 2004). The best-known and simplest approach to manufacture fullerene is the C_{60} fullerene consisting of 60 carbon atoms that form a sphere with a diameter of 0.7 nm. The atoms are arranged in pentagons and hexagons similar to a football. In future these fullerenes can be used, for example, in organic solar cells (Boeing, 2005). Meanwhile there exist several other even-numbered fullerenes in a range between C_{36} and C_{100} next to C_{60} fullerenes. In combination with metals of the first and second element groups, fullerenes can function as superconductors. For example, a compound of rubidium and a C_{60} fullerene conduct electricity without resistance at temperatures beneath 28 K (Binnenwies *et al.*, 2003).

Nanoparticles are of great interest as composite materials. Materials based on nanoparticles can be available as nanopowders, as clusters of compounded solid bodies, as nanocomposites, as agglomerates, or as nanostructures on solid-body surfaces. Through the deposition of nanocrystals, for example, on a surface, nano-islands can be manufactured in a way that they are qualified as catalysts (Paschen *et al.*, 2004).

2.2.1.2 Line-shaped structures

Nanotubes, nanowires, nanorods, or nanotrenches on surfaces are examples of line-shaped structures.

So far the most intensively examined structures are nanotubes made from carbon (CNTs). They consist of cylinder-shaped graphite layers and are structurally closely related to fullerenes.

In 1991 nanotubes were discovered by a coincidence when a Japanese scientist Sumio Iijima researched a new process to manufacture fullerenes (Binnenwies *et al.*, 2003). The diameter of nanotubes ranges from 1 nm up to 100 nm, and their length can add up to several hundred micrometres. CNTs can be manufactured in large amounts (ton scale) by using the manufacturing method of catalytic chemical vapour deposition (CCVD). In this method hydrocarbons or carbon monoxide are decomposed under high temperature with the help of catalyst particles. On their surface CNTs grow. It is possible to manufacture single-walled and multi-walled nanotubes. For the time being, a multitude of possible applications of CNTs is discussed. On account of their mechanical properties (e.g., high stability and tensile strength) they can be used as fibres for polymer composite materials and are classified as candidates for non-axial friction bearings (so-called ball bearings), respectively, for rigid and low-wear protective coatings. Based on their chemical properties, possibilities to store hydrogen are re-searched and discussed. Furthermore nanotubes could be employed as catalyst carriers, molecular filters, or membranes, as well as nanoreactors (Paschen *et al.*, 2004).

Besides nanotubes, nanorods pertain to line-shaped structures, too. Nanorods have relatively thicker and shorter rods with a diameter of up to a few ten nanometres as well as with a fourfold to tenfold length in relation to their diameter. Given that dimensions and geometry influence the band structure, research efforts focus then on tailoring the electric and optical properties. Possible applications for CdS nanorods are for photovoltaics in highly flexible plastic fabrics (Grüne *et al.*, 2005).

2.2.1.3 Layer structures

One of the major and broad industrial applications of nanotech-nology is layer structures. They are employed in almost every future technology ranging from microelectronic and optics to sensor systems, medicine, and wear-resistant coatings (BMBF, 2002).

These structures in nanoscale allow specific effects (e.g., quantum effects) or are preconditions to employ the specific properties of the layer material. To some extent, corresponding products are already on the market, for example, computer hard drives. Other examples are antibacterial, hydrophobic, and oleophobic textile coatings, which are on transfer from research and development to application. In the field of computer hard drives, diverse nanotechnological layer structures are already in use. The high-storage capacity of the so-called giant magnetoresistance (GMR) is based on a layer system consisting of iron layers that are separated by chrome layers with a thickness of a few nanometres. Other layer structures of great technical importance are ultrathin organic layers made from solid substrates. They are essential for the modification of surface properties such as wettability, sliding properties, and biocompatibility. Fields of application are in catalysis and sensor technology or as functional materials such as light-emitting polymer layers (Paschen *et al.*, 2004). In the field of energy technology nanolayers play an important role, too, because herein specific properties such as a high surface-to-volume ratio, high selective reactivity, or the ability for self-assembly can be used. Furthermore nanolayers have the benefits of resources preservation and cost savings if expensive materials (e.g., platinum) are used. Examples are fuel cells and catalysts (RS, 2004). Next to already available nanolayers, another research area is organic LEDs (OLEDs) that are based on thin-film systems (BMBF, 2002).

2.2.1.4 Pore structures

Materials with defined porosity are of great scientific and technical interest because interactions with atoms, ions, or molecules can occur on the surface as well as on the whole volume of the material. Surfaces of several square metres per gram are possible. Possible fields of application are catalysis, filtration, gas separation, heat insulation, and anti-reflective coatings. As a rule, in order to have an efficient and technical application, pore structures have to show a regular distribution of pores across the manufactured work piece.

Another possibility is the chemical modification of the surface of the pores, for example, by thin layers of catalyst material such as platinum or rhodium. Pore structures show also, next to the above-described functional aspects, advantages compared to solid materials in structural applications, for example, for lightweight constructions. One manufacturing method for pore structures is the sol-gel method. Therewith high porous oxidic glass and ceramics can be synthesised (via so-called aerogels) with densities lower than 0.01 g/cm^3 (Paschen *et al.*, 2004). Pore structures can be easily manufactured on the basis of a variety of materials. In industry, zeolites are widely used (Grüne *et al.*, 2005). Zeolites are naturally occurring minerals – crystalline aluminosilicates with a very huge surface – and they can be employed as catalysts in the petrochemical industry (Edwards, 2006). Others are suitable as adsorption agents for small molecules (e.g., water), for gas separation (e.g., nitrogen and oxygen), or as catalysts (Binnenwies *et al.*, 2003). Nanoporous pore structures are also employed for heat insulation and for anti-reflective coatings (Paschen *et al.*, 2004).

2.2.1.5 Complex structures

By clustering of molecular components, for example, supramolecular structures can be obtained from hydrogen bonds. Under qualified conditions construction of complex structures can often occur spontaneously and without exterior control. From a nanotechnological viewpoint, for example, liquid crystals and protein aggregations are important supramolecular structures.

Materials that own a crystal-like long-range order of their components but local mobility of single molecules are characterised as liquid crystal materials. Macroscopic material properties are controlled by a nanoscopic molecule structure. Current examples are displays or optical storage devices. Further applications are seen in the field of photoconductors, reflectors, and dyestuff. For photovoltaics as well as the field of medicine (drug delivery), complex structures are prophesied as a good future potential (Paschen *et al.*, 2004).

Branched polymers and dendrimers constitute another important material class. Thereby it is about structures that are generated on their own via a three-dimensional ramification. The molecule grows about 1 nm for every ramification from the inside out, like an onion (Edwards, 2006). Dendrimers show, based on their variable molecular architecture, multiple possibilities for designing various functionalities. It is, for example, possible to form different-sized excavations in the interior of the molecule and all areas of the macromolecule can be functionalised (Paschen *et al.*, 2004). Examples are liquid crystal grids, artificial antibodies, biosensors, drugs, and displays (OLED). Further future applications can be found in nanolithography, electronics, photonics, and chemical catalysis. Also in batteries or photovoltaic modules, dendrimers can play an important role in the future (Edwards, 2006).

2.2.2 Top-Down and Bottom-Up Strategy

According to (Paschen *et al.*, 2004) there are two strategies for further development at nanometre scale (see Figure 2.1). The so-called top-down strategy, which dominates physics and physical technology, and the bottom-up strategy that so far is

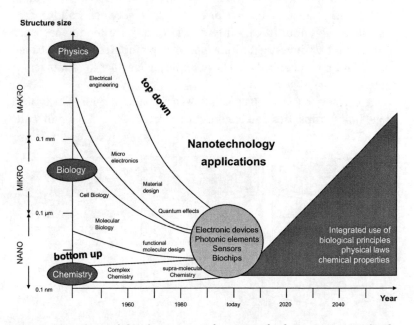

Figure 2.1 General development tendencies and relation to nanotechnology, based on (Bachmann, 1998).

represented by chemistry and biology. The top-down strategy means to miniaturise components and structures over and over based on microtechnology. The main driver of this approach is the electronic industry. In contrast to the top-down strategy, the bottom-up strategy concerns the atoms or molecules which build up nanostructures. Nature uses the bottom-up strategy to build up complex biological nanostructures, for example, functional biomolecules or cell organelle.

'Atoms on the small scale behave like nothing on a large scale, for they satisfy the laws of quantum mechanics. [...] At atomic level, we have new kinds of forces and new kinds of possibilities, new kinds of effects' (Feynman, 1960). A macroscopic body can be extensively described by the laws of classical physics. On the transition to nanoscale, there occurs a change to an object that can be described evermore by quantum mechanics. Moreover surface and interference layer properties play a more and more important role in comparison with volume properties (Paschen *et al.*, 2004).

Nanotechnology takes place in a threshold between individual atoms and molecules on the one hand and between solid bodies on the other hand. In this intermediate range, phenomena occur that cannot be investigated on macroscopic bodies and are based on quantum mechanical effects, an enlarged surface, and molecular detection (see Figure 2.2).

According to (Paschen *et al.*, 2004) a nanoparticle loses its typical solid body properties and can be rather regarded as a big molecule.

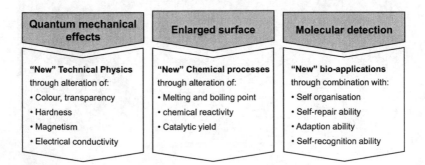

Figure 2.2 Modification of properties in the nanoworld (Luther *et al.*, 2004).

The electronic, chemical, and optical properties of a particle often change dramatically with its particle size. An integral basis of a number of properties of nanomaterials is based on the high ratio between surface and volume. In general the atoms on the surface have non-saturated connections that can have high reactivity and thereby influence the chemical properties (e.g., catalytic activity).

By an increasing surface ratio, the surface energy of the single particle increases when, for example, the melting point lowers or sinter activity rises. Materials and substances in nanoscale are a fundamental basis of nanotechnology because they show remarkable properties that cannot be found in conventional materials. To these belong, for example, super-plasticity, increased hardness, reduced or increased heat conductivity, and increased electronic resistance. In Table 2.2 some examples of effects and changes of properties of nanomaterials and components are listed.

2.2.3 Tools and Production Processes

The availability of preparative and analytic procedures that facilitates the manufacturing of a component, a device, or a basic material in adequate quantity and the control of its quality is significant for an industrial manufacturing of the described basic structures. The industrial aspect of nanotechnology is today, to a great extent, represented by the functionalisation of material surfaces. A synonym for this is, for example, the lotus effect. In the field of functionalisation of surfaces, partly long-known strategies are employed. The germ-killing effect of silver particles, for instance, has been well known for a long time. The innovative character of nanotechnology appears in the application of nanoscale particles in the form of novel composite materials, whereby their components are well known but their combination to an entire material is innovative. Top-down and bottom-up strategies will increasingly be adopted simultaneously to manufacture nanostructures and nanomaterials, whereas bottom-up strategies are deployed to manufacture key components such as fullerenes, polymers, or supramolecular structures that are assembled with further structures to entire components. The connection of the components to the macroscopic environment subsequently takes place with top-

Table 2.2   Examples for changes of properties of nanomaterials (Luther, 2007b; Luther, 2008)

Material property	Example for effects based on nanoscale configuration
Chemical	Easy-to-clean properties of surfaces by nano-particulate coatings
	More efficient catalysts for cars through extended surfaces and specific catalyst design
	More powerful batteries and accumulators through higher specific electrode surface
Mechanical	Increased gas tightness of food packages through admixture of nanoparticles or through nanocoatings
	Improved scratch resistance of coatings through ceramic nanoparticle
	Wear-resistant und highly resilient mechanical component through nanostructured layers
Optical	Optimised light absorption properties of solar cells through quantum dots and nanolayers
	Luminescent polymers for the production of energy-efficient organic light diodes
	Anti-reflection properties for solar cells to increase energy yield
Electronic	Electric insulators through nano-structured fillers in components of high-voltage power lines
	Enhanced thermoelectrica for more efficient power generation from heat through nano-structured layer systems
	Optimised electron conductivity through CNTs and nano-structured superconductors
Thermal	Nano-structured heat protection layers for turbine blades
	Improved heat conductivity of heat exchangers through CNTs
	Nanofoams as super-insulation systems in buildings due to the nanoporous structure

down strategies. In the long term, new production methods are essential for supramolecular strategies development, while the so-called mix-and-match-approaches between top-down and bottom-up strategies represent an important basis for structural elements (Hartmann, 2006).

Since a variety of technologies are already in applications or else in further development, this chapter describes merely examples of key production process methods of nanoanalytics. For continuative literature, please refer to Grüne *et al.*, (2005). Nanoparticles and nanostructures with at least one dimension in nanoscale can be manufactured of all three phasing. As a rule the processes are well-known proceedings that were optimised in terms of control, composition, and pattering of nanoscale materials (Paschen *et al.*, 2004).

In the scope of this book the most important process technologies are described. They are divided into five categories:

1. Vapour deposition
2. Manufacturing from liquid or dissolved raw materials
3. Manufacturing from solid raw materials
4. Lithography
5. Self-organisation

The process technology lithography is counted among top-down strategies. The remaining four process technologies to manufacture nanostructures are, by contrast, the bottom-up strategies. Tables 2.3 and 2.4 summarise the characteristics and attributes of top-down and bottom-up strategies as well as evaluate both strategies. Important examples are described in detail next.

Table 2.3   Characteristics of top-down and bottom-up proceedings (Haas and Heubach, 2007)

	Characteristics	Advantages, disadvantages, and examples
Top-down proceedings		
Lithography	Microelectronic process: structured hardening combined with etching processes	Microelectronics, micro-optics, resolution limited by wavelength of light, very high complexity for structures < 100 nm
High-energy milling	Grinding of conventional powders, high energy demand, upscaling possible	Problem: impurities, often used for metal carbides/silicides
Electrospinning	Industrial facilities available	Polymers
Bottom-up proceedings		
Sol-gel	Facilities cost efficient, small particles with close size distribution only in small quantities possible	Raw materials often relatively expensive
CVD	High demand for technical equipment, high demand on purity, small throughput	Relatively expensive, is used for CNT or fullerenes
Hydrothermal proceeding	Inexpensive raw material, great quantities possible	Limited material range

Table 2.4 Attributes and evaluation of top-down and bottom-up proceedings (+: low, ++: medium, +++: high) (Haas and Heubach, 2007)

	Nanoparticles	Nanofibres	Moulded bodies	Nanocomposites	Nanolayers	Costs	Application maturity	Energy demand
Top-down								
Lithography					•	++	+++	++
High-energy milling	•					+++	+++	+++
Highly plastic deformation	•		•			+++	++	+++
Electrospinning		•				++	+	+
Electronic arc discharge		•			•	++	++	++
Delaminating of clays and phyllosilicates	•			•	•	++	+++	++
Bottom-up								
Chemical precipitation	•					+	+++	+
Sol-gel	•		•	•	•	++	++	+
Flame reactors	•				•	++	+++	++
Electrochemical deposition	•			•	•	+	+++	++
Hydrothermal proceedings	•					++	++	++
CVD	•				•	+++	++	+++

2.2.3.1 Vapour deposition

Vapour deposition is mainly used for manufacturing thin layers by applying gaseous raw materials. Nanopowder as well as quantum dots can be manufactured through this process technology, too (Steinfeldt *et al.*, 2004). Proceedings of vapour deposition are vitally important for industrial processes because they can often be operated continuously.

The term 'chemical vapour deposition' (CVD) contains all proceedings, that is, a raw material vapourises and afterwards precipitates on a substrate surface. The crucial layer matches the raw material, or a new and varied material can be formed by chemical reactions. As a rule CVD proceedings are conducted under vacuum conditions (Theodore and Kunz, 2005). Important CVD proceedings are, for example, thermal CVD, plasma-active CVD, photo-CVD, and catalytic CVD, which is employed increasingly to manufacture CNTs (Steinfeldt *et al.*, 2004).

In case of physical vapour deposition (PVD), solid raw material is transferred in the gaseous phase by physical exposure (e.g., thermal energy) under vacuum conditions, and a thin film is accumulated on the surface of a substrate by condensation. An advantage of this proceeding is the relatively low precipitation temperature whereby an extra huge and coated spectrum of materials can be employed. Sputtering, evaporation, ion cladding, and molecular jet epitaxy are counted among PVD proceedings (Grüne *et al.*, 2005).

Aerosol technologies are based on sputtering a liquid raw material in subtlest droplets within a gaseous medium. After an aerosol is inserted into a heated reactor zone, the solvent is burned or vapourised. Using aerosol technologies, highly pure materials, ultrafine porous and non-porous powder, or thin films on a substrate can be manufactured. The pyrosol proceeding is of great importance. Thereby the aerosol is decomposed by means of a bath gas next to a substrate to be coated. The aerosol condenses in the shape of product steam particles on the substrate surface (Grüne *et al.*, 2005).

In flame-assisted deposition nanoparticles are manufactured by decomposition of liquid or gaseous raw materials in a flame. The size of the particles and the crystal structure can be varied by changing the concentration or the reacting agent, flame temperature as well as dwell period, or the raw materials in the flame (Steinfeldt *et al.*, 2004). Flame-assisted deposition cannot be used to manufacture thin layers (Grüne *et al.*, 2005). On an industrial scale nanoscale soot, so-called carbon black, is produced using flame-assisted deposition (Degussa, 2004).

2.2.3.2 Manufacturing from liquid or dissolved raw materials

The sol-gel process is an important wet-chemical technique to manufacture different nanotechnological products such as powders, thin films, aerogels, and fibres. It offers the possibility to combine organic and inorganic components, that is, to produce organically modified products (Steinfeldt *et al.*, 2004).

The sol-gel process begins with the reaction of liquid components in the solution, which generates fairly great molecules or nanoparticles. The sol can be generated and subsequently transforms into a gel condition in two ways. The generated molecules develop further, or the sol thickens until a gel-like solution is formed. Through destabilisation of the sol or gels, it is possible to precipitate nanoparticles into a defined size.

A typical proceeding to produce nanoparticles is precipitation of agents out of solution, whereas the process of precipitation is triggered with the help of qualified substances. Thereby either the composition of the solution changes by what the substance to be precipitated gets antisoluble in the solution, or a new compound is generated whose solubility is lower than its concentration in the solution. The formation of nanoparticles proceeds from crystalline nucleus or amorphous primary particles to particle agglomerations. The size of the particles can be controlled by adding a reagent (Paschen *et al.*, 2004).

2.2.3.3 Manufacturing from solid raw materials

Nanoparticles can be manufactured directly from solid bodies by mechanical impact (milling). By mechanical milling in high-energy mills, nanoparticles can be manufactured on a high throughput rate. As a rule mainly powdered raw materials are dry- or wet-technically processed in a milling container with grinding elements. By the application of midget-grinding elements with diameters of 0.2 to 0.5 mm, particles with diameters between 40 and 100 nm can be attained (Grüne *et al.*, 2005).

It is possible to excite solid-state reactions in solid bodies that result in structural changes by systematic heat treatment with temperatures beneath the melting point. This process is called tempering. With this proceeding it is, for example, possible that amorphous metals or alloys that were manufactured by accelerated cooling of a melted mass develop a nanocrystalline structure by controlled annealing (Grüne *et al.*, 2005).

Structures in nansocale in metals and alloys can be generated by strong mechanical deformation, too. An example of such proceedings is the equal-channel-angular-extrusion (ECAE) method. Thereby a cylindrical or a cuboid-shaped sample is pressed along a channel around a corner. If this operation is iterated, multiple high-deformation grades can be achieved. The high-pressure torsion method uses thin, flat blanks that are twisted under high hydrostatic pressure. During the ductile deformation a high defect density is generated in the material by which subgrains are formed. With rising deformation the sizes of these grains decline (Grüne *et al.*, 2005).

2.2.3.4 Lithography

Optical lithography produces nanostructures by increasing miniaturisation, and it is the most frequently used proceeding to manufacture nanostructures, for example, for manufacturing microchips.

With the help of structured reticles, light and X-rays are projected onto a sample surface that is coated with a photoresist. The generated structures, resulting after the processing of the lacquer, can be transformed on a substrate with etching processes. The smallest possible structures depend on the wavelength of light. Manufacturing processes with ultraviolet light (wavelength 248 nm) have reached their limits; now F2 laserlight sources (wavelength 156 nm) and X-rays (wavelength 13.5 nm) are researched in order to further miniaturise nanostructures (Paschen *et al.*, 2004).

2.2.3.5 Self-organisation

In the proper sense, self-organisation is no technical proceeding but rather a molecular building principle after growth and structure formation processes in nature are executed. In this process, molecules, atoms, or particles join together to form functioning units. A pivotal characteristic is that the information to build up structures is stored in each single component. A proceeding for manufacturing self-organised structures that is already on the market is the self-organised monolayer. Ultrathin layers whose structure is forced by the arrangement of the substrate atoms, because the attached molecules get connected to the substrate atoms, can be manufactured with this proceeding (Steinfeldt *et al.*, 2004).

2.2.3.6 Nanoanalytics

Nanoanalytics summarises all methods and instruments that qualify for the analysis and quality control of objects in nanoscale. Therefore it is an indispensable part of nanotechnology. The development of the scanning tunnelling microscope (STM) opened the door to the nanoworld in 1981. STM enabled us for the first time to directly uncover atomic structures (Paschen *et al.*, 2004).

In nanoanalytics, scanning X-ray microscope (SXM) methods play an important role. With this microscope method, a substrate is scanned with a micro probe tip and the occurring physical and chemical interactions between the substrate and the tip are measured. Several microscope variants count to the X-ray microscope methods, such as atomic force microscopy (AFM), the STM, or scanning near-field optical microscopy (SNOM).

Bibliography

Bachmann, G. (1998). Innovationsschub aus dem Nanokosmos – Tech-nologieanalyse, VDI-Technologiezentrum Physikalische Technologien, Abteilung Zukünftige Technologien, Düsseldorf.

Binnenwies, M., Jäckl, M., and Willner, H. (2003). Allgemeine und Anorgan-ische Chemie, Spektrum Akademischer Verlag, Heidelberg, Berlin.

BMBF. (2002). Standortbestimmung: Nanotechnologie in Deutschland, Bundesministerium für Bildung und Forschung, Bonn.

Boeing, N. (2005). Nanotechnik, *Technol. Rev.*, 32–41.

Degussa. (2004). Winzige Rußpartikel machen Reifen leistungsfähiger: Tanz im magischen Dreieck, Degussa AG, heute Evonik Industries.

Edwards, S. A. (2006). The Nanotech Pioneers: Where Are They Taking Us?, Wiley, Weinheim.

Feynman R. P. (1960). There's Plenty of Room at the Bottom An Invitation to Enter a New Field of Physics, *Caltech Engineering and Science*, Nr. 23:5, S. 22–36.

Grüne, M., Kernchen, R., Kohlhoff, J., Kretschmer, T., Luther, W., Neupert, U., Notthoff, C., Reschke, R., Wessel, H., and Zach, H.-G. (2005). Nanotechnologie: Grundlagen und Anwendungen, Fraunhofer Institut für Naturwissenschaftlich-Technische Trendanalysen, Euskirchen, Stuttgart.

Haas, K.-H. and Heubach, D. (2007). NanoProduktion - Innovationspotenziale für hessische Unternehmen durch Nanotechnologien in Produktionsprozessen, HA Hessen Agentur GmbH, Wiesbaden.

Hartmann, U. (2006). Faszination Nanotechnologie, Spektrum Akademischer Verlag, München, Heidelberg.

Köhler, M. (2001). Nanotechnologie: Eine Einführung in die Nanostrukturtechnik, Weinheim u.a.

Luther, W. (2007b). Nanotechnologie als wirtschaftlicher Wachstumsmarkt, Chancen und Risiken aktueller Technologien, A. Gaszó, S. Greßler, and F. Schiemer, Springer Verlag, Wien New York.

Luther, W. (2008). Einsatz von Nanotechnologien im Energiesektor, HA Hessen Agentur GmbH, Wiesbaden.

Luther, W., Malanowski, N., Bachmann, G., Hoffknecht, A., Holtmannspötter, D., Zweck, A., Heimer, T., Sanders, H., Werner, M., Mietke, S., and Köhler, T. (2004). Nanotechnologie als wirtschaftlicher Wachstumsmarkt: Innovations- und Technikanalyse, Zukünftige Technologien Consulting der VDI Technologiezentrum GmbH, Düsseldorf.

Paschen, H., Coenen, C., Fleischer, T., Grünwald, R., Oertel, D., and Revermann, C. (2004). Nanotechnologie - Forschung, Entwicklung, Anwendung, Springer Verlag, Berlin,Heidelberg, New York.

RS. (2004). Nanoscience and nanotechnologies: opportunities and uncertainties, Royal Society, Royal Academy of Engineering, London.

Steinfeldt, M., Gleich, A. v., Petschow, U., Haum, R., Chudoba, T., and Haubold, S. (2004). Nachhaltigkeitseffekte durch Herstellung und

Anwendung nanotechnologischer Produkte, Institut für ökologische Wirtschaftsforschung (IÖW) gGmbH.

Theodore, L. and Kunz, R. G. (2005). Nanotechnology: Environmental Implications and Solutions, Wiley-Interscience, Hoboken, New Jersey.

2.3 Innovation and Economic Potential

Dr. Wolfgang Luther
VDI TZ GmbH, Germany

Nanotechnology has made a tremendous development in the past decades. Since the 1980s, when the term 'nanotechnology' was known only among experts and futurists, it has become a prominent international research trend and a key technology to address future global challenges. Nanotechnology has received much attention and funding from governments around the world. By the end of 2008 nearly $40 billion had been invested by governments in nanotechnology research. And in 2009 alone, global government funding of nanotechnologies is estimated to a sum of $9.8 billion, with the European Union (EU) and the United States in a leading position (Cientifica, 2009). The dynamic development of nanotechnology can be demonstrated, not only by a steady growth in public subsidies, the number of patents, and publications in the past years, but also by an increasing number of nanotechnological, optimised products in the world markets.

2.3.1 *Nanotechnology as a Cross-Cutting Innovation Field*

Up to now, an internationally accepted and harmonised definition of nanotechnology is still lacking, although much work has been spent in standardisation work in recent years. Therefore, in the context of the following chapter, nanotechnology is understood as a conglomerate of all methods and processes for the controlled manufacturing, analysis, and technical application of structures and materials with dimensions of 1 to 100 nm. In this nanometre-size range, drastic changes of material and component properties can occur, which are used for a targeted functional optimisation of technological components. This concerns nearly all disciplines of natural and engineering sciences.

- Physical material properties: Solid-state properties like electric conductivity, magnetism, fluorescence, mechanical strength, and hardness change fundamentally with the number and arrangement of the interacting atoms and molecules. Energy states in nanoclusters are quantised and do no longer follow the principles of classical solid-state physics. As a consequence novel phenomena and effects occur, such as the photoluminescence of nanostructured silicon or the well-known GMR effect. The latter effect describes a drastic change of the electric resistance in certain nanoscale magnetic layer stacks when external magnetic fields are applied and forms the basis for high-performance reading heads in miniaturised magnetic hard disk storage drives.
- Chemical material properties: In nanoscaled structures, the ratio of reactive surface to inert bulk atoms of solid matter is drastically increased. For example, certain nanoporous substances like metalorganic framework substances exhibit specific surface areas of more than thousand square metres per gram. Therefore, nanostructured materials have an enormous application potential in surface-controlled solid-state reactions, for example, in the field of catalysis, electrochemistry, material separation, or even hydrogen storage.
- Biological material properties: Nanostructures also play a vital role in biology, since nearly all biological processes are controlled by nanoscaled objects, like nucleic acids, proteins, and other cell constituents. Nanotechnology enables, on the one hand, the observation and clarification of life processes through nanoanalytic processes like, for example, high-resolution optical microscopy, and, on the other hand, new approaches in medical therapy (e.g., intelligent drug delivery systems), in regenerative medicine (improved implants or bone substitution materials), and in diagnostics (optimised *in vitro* rapid tests, contrast agents, etc.). Another promising

research field of nanotechnology is the technical utilisation of self-organisation processes, where on the basis of chemical interaction and molecular recognition mechanisms, individual molecules are assembled to larger units and technological components.

By utilising these nanospecific properties and effects, nanotechnology opens up new opportunities for an intelligent design of materials by combining desired material properties and adapting them specifically to particular technical purposes. This will not only result in a broad commercialisation of nanotechnology in various industrial branches but will also contribute to solving urgent global challenges in societal demand fields like energy, water, health care, communication, and mobility.

In many cases nanotechnology is a further development and refinement of scientific disciplines with a long tradition, such as microelectronics, surface and coating technologies, or colloidal sciences. However, the innovation potential of nanotechnology reaches far beyond the current approaches of so-called passive nanostructures like nanocomposites or nanocoatings and probably will evolve to active structures, which can change functionality in response to the environmental stimuli, subsequently to hierarchical ordered systems of active nanosystems (i.e., artificial organs) and could finally lead to molecular systems and machines, which can be created bottom up from molecules with designed functionalities (Roco, 2004). In the long term a synergistic combination of four major areas of science and technology, often referred to as 'convergence of technologies', is expected to occur. These comprise technological fields, each of which is currently progressing at a rapid rate, namely, nanotechnology, biotechnology and biomedicine, information technology, and cognitive sciences.

Figure 2.3 gives an overview on nanotechnology applications and products for different industrial branches. The spectrum of the development status of nanotechnological applications comprises long-established nanoproducts, developments in the current market realisation or prototype phase respectively, and long-term to visionary research approaches, the commercial realisation of which is not to be expected within the next 10 years.

Civil Security Technologies	• Chemical/Biological (C/B) decontamination systems on nanoparticle basis • Security tags on the basis of nanoparticles and nanopigments	• Protection systems or the basis of nanofluids reinforcing on pressure impact • Lab-on-Chip-Systems for C/B diagnostics • Electronic noses for the detection of C/B-substances	• Super-absorbing gels for the neutralization of radioactive residues • C/B-filter systems on the basis of nanocatalytic or nanostructured materials • Nanotube-reinforced protection systems	• Self-healing protection materials • Early diagnosis systems on the basis of cross-linked nanosensors/NEMS • Biomonitoring systems with integrated molecular diagnostics and medication
Construction Engineering	• Dirt-resistant coatings and paints • IR-reflecting nanolayers for heat-absorbing glasses • Photocatalytic coatings for roof tiles, awnings, PVC-profiles • Nano-based sealing coatings	• Antibacterial paints (nanosilver) • Multifunctional ceramic wallpapers • Fire-protected glasses and construction materials • Aerogel facades, vacuum insulation panels • Switchable glasses (electro-/photochrome)	• Nanoporous insulation foams • Large-area flexible solar cells as facade elements • OLED-illumination • Ultra high-performance concrete • Nanooptimized asphalt mixtures	• Ultra-light construction material on CNT-basis • Multifunctional adaptive facade elements (energy recovery, shading, illumination) • Construction material with self-repair mechanism
Energy	• Nanostructured catalysts • Nanolayers for corrosion and wear protection • Nanomembranes for wastewater treatment • Anti-reflection layers for solar cells	• Nanooptimized micro fuel cells/batteries • Photocatalytic air and wastewater treatment with nano-TiO$_2$ • Heat-protection for efficient turbines • Groundwater sanitation with iron-nanoparticles	• Large-area polymer solar cells • Nanosensorics for environmental monitoring • Thermo-electric waste heat utilization • Efficient hydrogen generation through nanocatalysts • Selective pollutant separation	• Artificial photosynthesis • High-efficient quantum dot solar cells • Resource saving production through self-organisation • Efficient power supply lines with CNT-cables
Environment				
Textile	• Dirt-repellent textiles through nanoparticles • Antibacterial textiles through nanosilver • Scent-impregnated textiles on the basis of nanocontainers (e.g. cyclodextrine)	• UV-protected textiles through nano-TiO$_2$ • Thermal protection clothing with aerogels • Abrasion-resistant fibers through ceramic nanoparticles • Nanooptimized technical textiles	• Active thermal regulation through phase-change materials • Electrically conductive textile fibers for Smart Textiles, electrostatics etc. • Textile-integrated OLED • Textile-integrated power generation (e.g. thermoelectrics, solar)	• Textile-integrated sensorics/actorics for active movement support, control of body functions etc. • Textile-integrated digital assistance systems (Human Interfaces)
Automotive Engineering	• Nanostructured exhaust catalysts • Nanocoated Diesel-injectors • Anti-reflection layers for displays • Nanostructured admixtures for tires • Magneto-resistant sensors • Scratch-resistant clear lacquers, effect lacquers	• Nanoparticles as Diesel-additives • LED-headlights • Nanohard layers for polymer disks • Nanostructured light-construction composites	• Thin-film solar cells for car roofs • Nanooptimized fuel cells • Thermoelectric waste-heat utilization • Ferrofluids for adaptive shock absorbers • Nanoadhesives in production	• Switchable, self-healing lacquers • Adaptive bodyshell for optimum air resistance • Intelligent drive assistance and traffic detection • Connected cars

Figure 2.3 *(Contd.)*

	Established nanoproducts (Years until commercialization)	Market entry (0 - 3 years)	Prototype (4 - 10 years)	Concept (> 10 years)
Chemistry	■ Nanopowder/dispersions (TiO₂, SiO₂, ...) ■ Nanostructured industrial carbon black ■ Nanostructured active agents and vitamins ■ Polymer dispersions ■ Effect pigments ■ Ferrofluids	■ Fullerenes, Carbon Nanotubes (CNT) ■ Nanopolymer composites ■ Organic semiconductors ■ Semiconductor quantum dots ■ Inorganic/organic hybrid composites ■ Dendrimers	■ Nanoporous foams ■ Switchable adhesives ■ Functionalized nanomembranes ■ Artificial spider silk ■ Electrospun nanofibers	■ Self-healing materials ■ Self-organizing complex materials/composites ■ Moleculare machines ■ Adaptive multifunctional materials
Optics	■ Nanolayers for scratch-proof plastic spectacle lenses ■ Ultra-precision optics for telescopes etc. ■ Anti-reflection layers for glass coatings ■ LED, diode lasers	■ Optical microscope with nano-resolution ■ Organic light-emitting diodes (OLED) ■ CNT-field emission displays ■ 2D-photonic crystals for light conductors	■ EUV lithography-optics ■ Quantum-dot lasers ■ Quantum cryptography ■ 3D photonic crystals	■ All-Optical Computing ■ Optical meta-materials for "Magic Cap Applications" ■ Data transmission through surface plasmons
Electronics	■ Hard-disk storage units with GMR-reading heads ■ Silicon electronics (structures < 100 nm) ■ Flash-storage ■ Polymer electronics e.g. for RFID-tags	■ Silicon-electronics 32 nm structures ■ CNT-field emission displays ■ MRAM-memories ■ Phase-change memory	■ MEMS-memory ("Millipede") ■ CNT-data memory ■ Silicon-electronics 22 nm structures ■ CNT-interconnects in circuits	■ Molecular electronics ■ Quantum computing ■ Spintronic logics ■ DNA-computing
Medicine	■ Nanoparticles as contrast medium in diagnostics ■ Nanoscale drug carriers ■ Biochips for in-vitro diagnostics ■ Nanomembranes for the dialysis	■ Nano cancer therapy (hyperthermy) ■ Nanostructured hydroxylapatite as bone substitute ■ Quantum-dot markers for diagnostics ■ Controlled drug release for implants	■ Bio-compatible, optimized implants ■ Nanoprobes and markers for molecular imaging/diagnostics ■ Selective drug carriers	■ Artificial organs through Tissue Engineering ■ Theranostics ■ Neuro-coupled electronics for man-machine-interfaces and active implants

Figure 2.3 Examples of application spectrum and time perspectives of nanotechnological developments (*Source:* VDI TZ).

2.3.2 *Economic Relevance of Nanotechnology*

Nanotechnology is often referred to as a key technology with great economic importance, but a quantifiable evaluation and a transparent illustration of the actual status quo of the economic implementation are difficult to achieve. One reason for that is the fact that nanotechnology as an 'enabling technology' sets in at a relatively early stage of the value-added chain, that is, at the manufacturing of nano-optimised materials, nanoscale coatings, or nanostructures in electronic circuits. In the end product it is often hardly noticeable for consumers if and how nanotechnology contributes to the performance of the product. Another reason for the rather diffuse external perception of nanotechnology is the abundance of many different processes and application fields, which render the definition of nanotechnology difficult. The broad application range of nanotechnology includes not only high-tech fields, like electronics, optics, life sciences, and new materials, but also more traditional branches, like mechanical engineering, the textile and construction industries, or even products of daily use, like cosmetics, sport equipment, and household products. International standard-setting bodies, such as the International Organization of Standardization (ISO) and International Electrotechnical Commission (IEC), are currently developing the basis for a harmonised nomenclature and standardisation of nanoscale objects and processes to achieve an internationally harmonised view and definition of nanotechnology (Bard, 2009). Although the first set of nanotechnology standards has meanwhile been implemented, this procedure will still take some years, and it is a moot question whether it is possible to define all facets of nanotechnology in a standardised framework and to clearly dissociate them from adjacent disciplines like microtechnology or chemistry and materials technology.

Today, nanotechnology already is an indispensable factor for the economic competitiveness in many branches of economy – in particular in the mass markets of electronics, chemistry, and the optical industry. Hard disk drives, computer chips, or ultraprecision optics are only a few examples of products which depend inevitably on the manufacturing of nanostructures.

Here, the application of nanotechnological processes (e.g., lithography or coating methods for the manufacturing of nanostructures) is imperative and has already been practised and further developed for years. In other fields of application, nanotechnology offers an added value for the customer through additional high-quality functions and performance characteristics not achievable through conventional processes. Examples of this are nano-optimised components in automobiles, such as wear-free motor components, scratch-resistant lacquers, efficient exhaust gas catalysts, or rolling-resistance-optimised rubber mixtures in tyres. In the medium to the long term, nanotechnologies will also gain significant commercial influence in the fields of traditional industrial branches such as construction and textile industries.

The difficulty with regard to the exact determination of the market potential of nanotechnology is the fact that individual nanoscaled components, like nanolayers or nanograins in a nancomposite, can hardly be monetarily quantified, because these intermediates are generated *in situ* in the production process and are not directly sold in the market. In addition, nanotechnology often provides for the optimisation of individual components of a product,, for example, through more efficient materials or the utilisation of nanoscale effects, which are decisive for the competitiveness of the product but rather neglectable for the share in the total value of the product. The same applies to their influence on the unique selling propositions of optimised products as well as on the added value resulting for the customer. With the simply determinable market value of the entire product (e.g., of an automobile or a computer), being considered a 'nanoproduct', the economic relevance of nanotechnology is certainly overvalued. On the other hand, a mere summation of market values of individual nanomaterials and components would lead to an undervaluation of the economic relevance of nanotechnology, since its leverage effect as an 'enabling technology' would be left unconsidered (cf. FMRE, 2009).

To assess the economic potential of nanotechnology with a realistic view, it is necessary to apply a differentiated approach, which, on the one hand, quantifies concrete market segments on the value-added stage of marketable nanomaterials and tools/equipment for the manufacturing of nanostructures and, on the other hand, describes the leverage effect on the market volumes of nanotech-nologically influenced components and applications. The economic leverage effect of nanotechnology has been estimated by different market research institutes such as Lux Research, which predicts an increase in the world market volume from $147 bn in 2007 up to approx. $3 trillion in the year 2015 for the value-added stage of nano-optimised products, at an average compound annual growth rate (CAGR) of 46% (Lux, 2008). Here, nanotechnology is said to have the greatest influence on the fields of materials and production technology (increase from 97 bn in 2007 to 1,700 bn by 2015), followed by the electronics sector (inter alia, semiconductors, displays, batteries) with a growth from $35 bn in 2007 up to $970 bn by 2015, as well as on the health sector (pharmaceuticals, medical engineering, and diagnostics) with a growth from $15 bn in 2007 to $310 bn by 2015. The market potential of nanotechnology in the year 2015 estimated by Lux Research would correspond to approx. 5% of the current world gross domestic product or to 15% of the industrial goods market, which would mean that a large part of the global production of goods,, for example, in the fields of health, information, and communications technology, as well as in energy and environmental engineering, would be based on the application of nanotechnological know-how. Other assessments for the period mentioned assume world market potentials of nanotechnologically optimised products in the range of $1 to $1.5 bn (cf. e.g., NSF, 2001).

Certainly, these forecasts can only be regarded as rough overall estimations for the assessment of the economic leverage effect of nanotechnology, since it is not clear which components precisely are included in the assessment and with which market volumes. Market-oriented potential analyses require considerably more detailed and more differentiated breakdowns according to the respective market segments.

The main market segments of nanotechnology can be differentiated into nanoanalytics (equipment for the characterisation of nanostructures), nanotools (equipment for the manufacturing of nanostructures), and nanomaterials (materials and composites comprising nano-objects like nanoparticles, fibres, etc.). Due to the variety of different technological approaches and processes, a market assessment can hardly be complete. Nevertheless, there are some market forecasts, for example, by BCC Research, comprising at least the main product groups and technologies (cf. BCC, 2007a,b,c,d,e). In the field of nanoanalytics (electron, ion, and scanning probe microscopy and optical microscopy and equipment), a moderate growth from $2 bn in 2008 to $3 bn by 2010 is forecasted. The market segment of nanotools (CVD, PVD, lithography, and nanopositioning systems) is assessed at $35 bn for 2008 and at $43 bn by 2010. In case of nanomaterials, which comprise nanoparticles, nanofibres, and nanocomposite materials, a market volume of $2.1 bn is estimated by 2010 (cf. Figure 2.4).

These market assessments provide a realistic but also incomplete picture of nanotechnology potentials, since only a fraction of the multitude of different methods and material classes is considered. A more detailed assessment requires the breakdown to individual material classes and methods. The dynamic growth in some areas of nanotechnology, in particular in the field of new nanomaterials, such as CNTs, graphene, or quantum dots, double-digit annual growth rates are forecasted for the next years, which raises expectations for a rapid development and dissemination of new nano-based components and products.

2.3.3 Nanotechnology Companies in the Value-Added Chain

Companies in the nanotechnology field can be found along the whole value-added chain and comprise manufacturers in the field of nanomaterials, nanotools, nanoanalytics, and equipment for the operation of nanotools (e.g., vacuum and clean-room technology, plasma sources, sputtering targets, vibrational control devices, etc.) manufacturers and users of nano-optimised components and systems as well as service providers in the field of consulting, contract coating, technology transfer, contract analysis, and research.

Numerous companies are specialised in the marketing of equipment and contract services in the area of nanocoating and surface finishing of different materials and components. With nanocoatings, the functionality of nearly all material classes can be improved – often even several functions can be realised at the same time. Also well presented are lithography methods, like

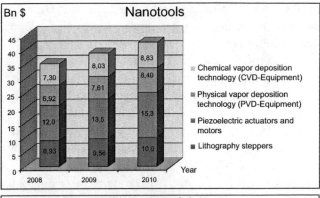

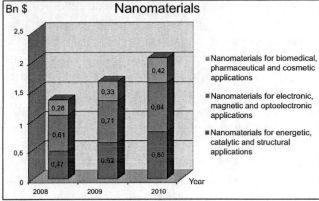

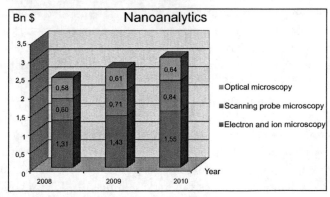

Figure 2.4 Forecasts on the development of the market potentials in different nanotechnology fields (cf. ref. FMRE, 2009). See also Colour Insert.

optical lithography, particle beam lithography, and nanoimprint lithography, which are utilised to impress nanostructures, with the highest precision, onto substrates, predominantly for components in electronics and optics.

The ultraprecise surface treatment by means of grinding and polishing machines or through laser/plasma treatment is another focal point of nanotechnology companies. In the field of nanoanalytics, atomic force microscopes and variants deriving from them are represented above average but include many manufacturers in different market segments of very diversified nanoanalytical tools such as electron microscopy, optical surfaces, and particle-measuring devices (e.g., confocal laser microscopy and laser scattering methods), which enable the characterisation of surface topographies or particle sizes with nanoresolution. The manufacturing of nanomaterials concerns mainly nano-structured metal oxides, polymer composites, and organic-inorganic hybrid materials. There are also a significant number of companies marketing nanotechnology products for consumer applications like cosmetics, clothing, personal care, and sporting equipment.

Some indication about this can be gained from a recent nanotechnology product inventory by the Project on Emerging Nanotechnologies at the Woodrow Wilson International Centre for Scholars in the United States (Nanotech, 2010), listing more than 1,000 nanotechnology consumer products.

The whole number of companies worldwide in the nanotechnology sector is difficult to estimate. Various nanotechnology business registers have been compiled with up to around 1,700 entries worldwide but with the shortcoming that no internationally accepted definition of such companies exists. The biggest share of industrial nanotechnology activities and companies is assumed to be in the United States, followed by Asia and Europe. Start-up companies are easier to identify as a nanocompanies because they are more eager to advertise their nanotechnology involvement for promotional reasons. Depending on the definition the number of nanotechnology companies can vary significantly, for example, the figure is estimated at 800 nanotech companies in Germany alone according to a recent national survey (Nanomap, 2010). Also the number of nanotechnology-related jobs is also difficult to assess.

The forecast ranges from 2 to 10 million new workers in 2015 worldwide. The largest number of nanotechnology companies can be assigned to SME with a significant share of innovative start-up companies, which are very important in the innovation process and the commercialisation of nanotechnology, because of their flexibility and their readiness to promote new technological developments with a high financial risk. However, the time span to bring a nanotechnology invention to the market as a product can be quite long and often takes more than 10 years, so some start-ups in the nanotechnology still were lacking commercial success.

Bibliography

Bard, D. Mark, D., and Moehlmann, C. (2009). Current standardisation for nanotechnology, J. Phys.: Conference Series, 170(2009), 012036.

BCC, (2007a). "Nanostructured Materials for Energy, Catalysis & Structural Applications", BCC Report December 2007 (http://www.bccresearch.com/report/NAN017E.html)

BCC, (2007b). "Nanostructured Materials: Electronic/Magnetic/Opto-electronic", BCC Report December 2007 (http://www.bccresearch.com/report/NAN017F.html)

BCC, (2007c). "Nanostructured Materials for the Biomedical, Pharmaceutical, & Cosmetic Markets", BCC Report November 2007 (http://www.bccresearch.com/report/NAN017D.html)

BCC, (2007d). "Physical Vapor Deposition (PVD): Global Markets, BCC Report October 2007 (http://www.bccresearch.com/report/MFG015C.html)

BCC, (2007e). "Microscopy: The Global Market", BCC Report June 2007, (http://www.bccresearch.com/report/IAS017B.html)

Cientifica. (2009). Nanotechnology Takes a Deep Breath ... and Prepares to Save the World., Global Nanotechnology Funding in 2009, white paper, Cientifica Ltd., London, April 2009.

FMRE, Federal Ministry of Research and Education. (2009). Nano. DE-Report 2009 – Status Quo of Nanotechnology in Germany, Bonn, Berlin, 2009

Lux, Research. (2008). Nanomaterials State of the Market Q3 2008: Stealth Success, Broad Impact. Available at HYPERLINK "http://www.luxresearchinc.com" www.luxresearchinc.com

Nanotech. (2010). Woodrow Wilson International Center for Scholars, Project on Emerging Nanotechnologies. Available at HYPERLINK "http://www.nanotechproject.org" www.nanotechproject.org.

Nanomap. (2010). Competency Map Nanotechnology in Germany. Available at www.nano-map.deNSF. (2001). Societal Implications of Nanoscience and Nanotechnology, Report of the National Science Foundation, March 2001. Available at www.nsf.gov.

Roco, M. (2004). Nanoscale science and engineering. Unifying and transforming tools, AIChE J., 50(5).

2.4 Risk and Safety Issues

Niels Boeing
Journalist, Germany

When nanotechnology makes headlines in the mainstream media nowadays, it is more often than not a variation of 'small particles, big risk' (Schmundt and Verbeet, 2008). Though it is one of the duties of the media to pay close attention to new technologies' downsides, this notion is ambivalent. Not only does it equate nanotechnology quite often with more mundane applications such as sunscreens or coatings that are employing nanoparticles for a useful effect, but it is also far from being representative of the potential of nanotechnologies. It also suggests that dealing carefully with nanoparticles is all that is at stake. So this notion underestimates at the same time both the benefits and the risks of nanotechnologies, which is a rather deplorable state of affairs.

In this article I will first show how this image of nanotechnology evolved in the interplay of the media, science, and politics and then propose a systematic approach to what I call primary and secondary risks of nanotechnologies.

2.4.1 *The Image of Nanotechnology: Three Phases*

From the first isolated mentioning of the term 'nanotechnology' in a paper by Norio Taniguchi in 1974 (Taniguchi, 1974) its meaning has changed markedly. Reflecting on how science, politics and media depicted the concept I distinguish three phases of the image of nanotechnology.

2.4.1.1 Pre-2000: the futuristic phase

It was in the 1990s that the mention of nanotechnology in the public media started to increase (Wolfe, 2003). At that time it was regarded as a powerful and visionary technology that one day would enable us to assemble and repair things at the nanoscale.

Typical illustrations showed nanomachine parts made from a few atoms, medical nanobots roaming the bloodstream, or space elevator cables made of CNTs (GermanMedia, 1997–2000).

In this futuristic phase the nanotechnology concept of Eric Drexler as drafted in his books *Engines of Creation* (1986) and *Nanosystems* (1992) had left a mark in the imagination of the public that was not lost on the scientific community to use for promotion of new approaches to material science and engineering. Moreover, though Drexler had not invented the term 'nanotechnology' in the first place, it was he who popularised it with his idea of an 'assembler' technology using mechanosynthesis for putting atoms together like LEGO bricks (Schummer, 2009). The climax of this phase can be seen in the US National Nanotechnology Initiative (NNI) of 2000 when its preparatory NSTF report borrowed directly from Drexler's idea with its title 'Shaping the World Atom by Atom' (Roco, 2003).

2.4.1.2 2000–2006: the nanomarkets phase

But it was also the NNI that prepared the shift to the second phase in which the economic potential of nanotechnology came into focus. Instead of portraying nanotechnology as potentially making age-old dreams like the land of milk and honey come true, it was now perceived as paving the way for huge, new markets. Emblematic was the vague forecast of a trillion dollar market for nanotech-enabled products in 2015 that Bainbridge and Roco made in a report for the US National Science Foundation in 2001 (Roco, 2001). The NNI had also helped broaden the term to everything technical below the threshold of 100 nm (NSTC, 2000) so that many projects and concepts in science and engineering could claim to be part of this new technology.

In a historical perspective this made sense because a broad range of disciplines had contributed to exploit the nanoscale technically in an unprecedented way, some of them already for decades. However, the notion that nanotechnology was a more or less coherent technology that dealt with 'nano' objects was not challenged.

At the same time the renunciation of a Drexlerian sense of nanotechnology had also been furthered by Bill Joy's gloomy 'Wired' essay (Joy, 2000) that took Drexler's own warning of nanobots getting out of control seriously, thereby kicking off the first debate about nanorisks. It was outright dismissed as a far-off doomsday scenario that did not fit the new optimism, captured, for instance, by a *Scientific American* special issue that announced 'Nanotech – The Science of the Small Gets Down to Business' (SA, 2000). The 'real' nanotechnology suddenly appeared as a godsend to initiate the next boom after the new economy of the nineties had spectacularly crashed. Several rating and investment companies set up nanotech stock indices (Merrill, 2004). Though some promises of nanotechnology were still as big as before like the end to cancer by 2015 (NCI 2003), thanks to nanomedicine, or supercomputers for everybody, thanks to nano-electronics, they mostly lacked any science fiction overtones now. Nano had finally become cool, when in 2005 the computer manufacturer Apple named the small version of its famous MP3 player, the 'iPod nano.'

2.4.1.3 Since 2006: the sceptical phase

But this situation was not to last. A report in early 2003 by the relatively unknown Canadian civil rights organisation ETC Group about the dangers of nanoparticles (ETC, 2002) and some early hints from toxicologists that, for instance, CNTs might become a new asbestos (Brown, 2002) had made a first negative splash in the media but not yet a wave. This changed, ironically, with a false alarm that triggered the third, sceptical phase. In April 2006 a bathroom sealing spray called Magic Nano had to be withdrawn from the German market after dozens of customers had complained about respiratory problems after its use. Eventually it turned out that the adverse health effect was not due to nanoparticles the seller claimed to give the product its extra value but from ordinary aerosols – in fact there were no nanoparticles found at all in the spray (BFR, 2006).

Nevertheless the international media reaction was quite strong, and the *Economist* asked, 'Has all the magic gone?' (TE, 2006) In a sense, yes, it had.

The incident came together with a growing number of toxicological findings that at least indicated that health risks from nanoparticles could not be excluded. It also left a bitter aftertaste that sticking the term 'nano' to products could as well be a marketing trick. The big promises now appeared not to materialise any time soon, while existing products labelled as nano – for instance, the ones listed in the nanoproduct inventory of the Woodrow Wilson Center (Woodrow Wilson, 2010) – seemed at odds with the nano hype before. Because this first generation of 'nano products' mainly employed chemical nanotechnologies, it allowed environmental organisations and groups to frame the associated risks along existing lines of older conflicts about chemical substances – in this case, nanoparticles.

Let us sum up this development. Drexler coined the term 'nanotechnology' as a special and distinct new technology that was later stripped of his too futuristic ideas. Recast as simply the next big and exciting technology after the Internet, the meaning of nanotechnology became both blurred and laden with vast economic expectations that have not yet been met by the first generation of nanotechnologies – the only ones somehow visible to the public. In this process the perception of nanotechnology has narrowed, and so has the perception of its risks.

2.4.2 *A Systematic Approach to Nanotechnology Risks*

As history shows, any new technology can have negative impacts on human health, the environment, society, or the economy that were not intended by inventors and innovators. Because of the scope of nanotechnology realised already in the futuristic phase, nanotechnology assessment set in quite early. Part of the motivation behind it was to prevent a public backlash like the one that happened with gene technology in the 1990s, which was attributed to poor communication to the public that did not take their concerns seriously. The US National Science Foundation issued a first broad assessment in 2001 (Roco, 2001), as did the Office for Technology Evaluation at the German Bundestag TAB in 2003 (Paschen *et al.*, 2003). The most highly regarded assessment eventually became the 2004 UK report of the Royal Society and Royal Academy of Engineering that also made some strong recommendations on handling nanomaterials (Royal, 2004). However, the influence of those reports remained limited to experts in academia, politics, the industry, and environmental organisations. Furthermore they did not provide a concept that would both be easily accessible and systematically classify different risks of nanotechnologies along risk dimensions and areas potentially affected by them. According to the focus of public attention and the urgency of dealing with risks as nanotech applications move from the lab to the market, I first distinguish two categories, primary and secondary nanorisks. Primary nanorisks are related to the potential impacts on human health and the environment, whereas secondary nanorisks could affect society and economic development by and large.

2.4.2.1 Primary nanorisks: impacts on health and the environment

According to risk research, risk can be defined as (Chicken and Posner, 1998):

$$risk = exposure \times hazard \tag{2.1}$$

Exposure means the probability with which a subject could come into contact with a technological object. Hazard stands for the potential damage a technological object could cause. Risk denotes then a probability for dangerous event. If this concept is applied to different nanotechnologies concerning their impact on health and the environment, three classes of primary risks can be identified.

2.4.2.1.1 Primary risk class 1: confined nanotechnologies A wide range of nanotechnologies rely on nanoscale components that are confined in closed systems and therefore cannot come into direct contact with organisms, neither the human body nor bacteria in the environment. Thus exposure is close to zero, reducing their risk close to zero as well – apart from the fact that for some nanocomponents, it's even hard to imagine a direct hazard at all. Some of the currently most matured nanotechnologies belong to this class, such as coatings and compounds, electronic devices, sensors, or scanning probe microscopes (SPMs). In coatings and compounds nanoparticles are embedded in a chemical matrix from which they could only be removed with a significant deployment of energy. Examples are polymers reinforced with CNTs as used in tennis rackets, steel reinforced with carbon nanoparticles (this use dating back to medieval times, for instance, in damascene swords – nature), and also the application of TiO_2 nanoparticles in Grätzel cells, a new kind of solar cell technology that enables flexible photovoltaics.

Modern computer hard drives exploit the effect of GMR, discovered by Albert Fert and Peter Grünberg in 1988, that works with ferromagnetic layers a few nanometres thick. This is a classical example of an effect that only occurs at the nanoscale. The same applies for SPMs the tips of which make use of the quantum mechanical effect of tunnelling. Some nanosensors employ tiny cantilevers that bind molecules for detection. The nanolayers in hard drives, the tips of SPMs, and cantilevers are confined in a device.

All of these examples are true nanotechnologies in the sense of the internationally agreed definition and some of them even in a more rigorous sense that they use particular nanoscale physical effects.

They have significant benefits for science, engineering, industry, and even everyday life but don't pose a grave primary risk for human health.

Ignoring them or even dismissing them as irrelevant for a risk assessment of nanotechnology as a whole – as it is sometimes done in the ongoing debate – reveals an ahistorical and blurred understanding of nanotechnology. However, it should be acknowledged that it is not yet clear if some of the nanocomponents of these technologies pose a risk after disposal.

2.4.2.1.2 Primary risk class 2: bioactive nanotechnologies Though risk class 1 might account for the lion's share of mature nanotechnologies, there is yet a small group that encompasses bioactive nanomaterials. That is, it has a direct impact on cells, either by intention or by circumstance. The first subset, 'by intention', comprises most of nanomedicine: nanoparticles are being developed as containers for more precise drug delivery or as direct therapeutic agents that kill tumour cells. An example of the latter are paramagnetic iron oxide nanoparticles that exert an effect on cells by overheating when put into an oscillating motion by an alternating magnetic field. Here exposure is high out of necessity because otherwise the medical effect could not be achieved. A potential hazard of these nanomaterials could arise from side effects that they have on tissue other than the target tissue they are designed for.

The second subset 'by circumstance' comprises nanomaterials in particle formulation that are not intended for interacting with cells but nevertheless might come into contact with them. In nanomedicine there are diagnostic agents, such as luminescent quantum dots, that are made of semiconductor cores (like gallium arsenide or cadmium selenide) surrounded by a layer of organic molecules.

Also part of this subset are free nanoparticles that could be taken up through inhalation, ingestion, the skin, or, in the case of microorganisms in the environment, the cell membrane. It is this subset of nanomaterials that is currently the focus of concern of nanotech critics.

These nanoparticles are temporarily in a loose form during production, like CNTs, thus posing a problem for safety at the workplace, or at the end of their life cycle, like silver nanoparticles ('nanosilver') washed out of textiles (McKenna, 2008), thus probably affecting microorganisms in sewage water and municipal sewage plants.

Also some nanoparticles are only loosely embedded in a suspension, such as zinc oxide or titanium dioxide nanoparticles in the notorious sunscreens, or adhered to fibres in functionalised textiles.

There is meanwhile some evidence that these nanoparticles, once in contact with cells, can have toxic effects and cause lesions and inflammation in tissue (Oberdörster, 2005; Nel, 2009). Especially CNTs have been shown to exhibit an asbestos-like behaviour, depending on whether their length exceeds a few micrometres and whether they come in tangled aggregates (Poland *et al.*, 2008). However, a complete assessment of the real hazards of free nanoparticles has not yet been achieved. One reason is that in some cases lesions could also be attributed to contamination of nanoparticles by impurities and catalytic materials from their production process (Wick *et al.*, 2007). A second reason is that in most studies negative effects only arose from a high dosage of the nanomaterial under scrutiny. Perhaps most important is the fact that the understanding of interactions at the nano-bio interface is only at the beginning. The most comprehensive overview to date was recently published by Nel *et al.* in *Nature Materials* (Oberdörster, 2005; Nel, 2009). They show that the effect of nanoparticles on cells depends on a range of factors such as the kind of interaction (e.g., Van der Waals, electrostatic, or electrosteric), the geometry of a particle, and the nature and chemical composition of the aqueous medium (e.g., water, blood serum, or interstitial fluid).

As a consequence it is impossible to generate a kind of formula that would allow us to assess the toxicity of a given nanoparticle *ex ante*.

However, the final report of the German research project Nanocare indicates that results from cell probes in the lab are quite consistent with results from animal experiments (Kuhlbusch *et al.*, 2009). So there is some confidence that high-throughput screenings with cell cultures could provide reliable results, thereby reducing the need for more controversial animal experiments in the future.

2.4.2.1.3 Primary risk class 3: autonomous nanosystems When nanotechnology first appeared as a future technology in the 1980s, it came along with the idea of powerful nanorobots for which Drexler coined the term 'assembler'. Their overall size as envisioned by Drexler was similar to that of cells. Also like cells, assemblers would be constituted of an intricate nanomachinery that would position atoms or molecules individually according to a construction plan (Drexler, 1986; Drexler, 1992). This would be stored and carried by an information processing unit, another feature shared between assemblers and cells: in assemblers a nanomechanical computer; in cells the chemical interplay of the genome and RNA. The fourth similarity was that assemblers would be able to replicate as do microorganisms. In his manifesto-like book *Engines* *of Creation*, Drexler already mused that the assemblers as more or less autonomously operating nanosystems could go astray and start to disassemble biological systems into their atomic and molecular constituents in order to get the raw materials necessary for further replication. This worst-case scenario came to be known as 'Grey Goo' (Freitas, 2000).

Fortunately it remains entirely hypothetical as there are not even prototypes for parts of the assembler technology on the horizon. However, Drexler emphasised that this technology would not be built from scratch but only emerge after a transition via new biological nanomachines that still employed conventional biological molecules. This intermediate step has come closer to reality with the development of synthetic biology. Although there are no references to the Drexlerian nanotechnology concept some the terminology of synthetic biology is taken from engineering (Endy, 2005).

Cells are depicted as a chassis in which gene circuits enable the production of substances or effects that don't occur in natural microorganisms. Among others the biotechnologist J. Craig Venter has repeatedly announced that his ultimate goal is to create synthetic bacteria that should serve as microfactories for biofuels (Venter, 2008).

As their molecular ingredients are certainly on the nanoscale, synthetic biology could be viewed as a kind of advanced bionanotechnology.

The community of synthetic biology has acknowledged from the outset that hazards are looming in its development even if not realistic to date (Maurer *et al.*, 2006; Garfinkel *et al.*, 2007). One uncertainty is how synthetic microorganisms could behave in the wild, though scientists assure that the associated risk is minimal because such microorganisms could not survive due to their unnatural specialisation that they got in the lab (Benner and Sismour, 2005). Of greater concern and taken more seriously is the possibility of biohacking: the very techniques that make synthetic microorganisms useful could be exploited to create new pathogens, perhaps even biological weapons (Maurer *et al.*, 2006; Garfinkel *et al.*, 2007; Benner and Sismour, 2005; Church, 2005).

2.4.2.2 Secondary nanorisks: impacts on society and the economy

The nanorisk debate so far has been focussed on the primary risks discussed before. Less emphasis has been put on adverse socioeconomical impacts that even safe and useful nanotechnologies could have. It is out of the scope of this article to analyse them in depth, so I will only briefly outline some of the problems. The main concern is that they could give rise to a 'nanodivide' between nations with advanced nanotech capabilities and others without, thereby further increasing the global development gap (ETC, 2005, Jones, 2007). Though more than 50 nations, among them newly industrialising countries, have set up a national nanotechnology policy like the NNI, a great deal of nanotech research takes place in the United States, the EU, Japan, and recently also China, India, and Russia.

These regions account for the majority of scientific publications as well as funding and patents (Lux, 2007). So the risk of a nanodivide is not merely hypothetical, although it is unclear how it will eventually manifest.

But even within industrialised countries nanotechnologies could at least support problematic trends. The emphasis of nanomedicine to finally enable a fully personalised medicine raises further questions about individual patients' data protection and discrimination.

Nordmann also points out that nanomedicine could probably intensify the development towards a reductionist and merely technological understanding of medicine that has dominated for the past decades (Nordmann, 2006).

Nanotechnologies for sensoring and computing could also facilitate more elaborate surveillance systems because of their potential to miniaturise relevant technologies in order to further intrude into the privacy of citizens. Moreover, Altmann has cautioned that nanotechnologies for military use could destabilise international security and perhaps even trigger another arms race in the coming decades (Altmann, 2006).

2.4.3 *Conclusion*

The change of perception that nanotechnology, as an umbrella term, has gone through has led to a blurring of its meaning. But it is important to stress, time and again, that nanotechnology is no coherent, distinct technology, and there is neither a corresponding coherent, distinct nanorisk. If anything, nanotechnology can be viewed as a new phase of technology in general with a huge potential for change, encompassing a multitude of concrete nanotechnologies. That said, it is the order of the day to differentiate them and the risks associated with them. Therefore I have proposed a classification of nanorisks which could help, first, to separate minor risks for health and the environment from serious ones, so that beneficial, but hardly hazardous nanotechnologies can be successfully implemented, and second, put the current focus on nanoparticle risks into perspective. For there are further risks, both for health and the environment and for society and economic development that are not yet quite tangible but potentially serious – probably more serious than nanoparticle risks.

Bibliography

Altmann, J. (2006). Military Nanotechnology: Potential Applications and Preventive Arms Control, Routledge, 2006.

Benner, S. and Sismour, M. (2005). Synthetic biology, Nat. Genet., 6, July 2005, 533–543; they write: 'The 30 years of experience with genetically altered organisms since Asilomar have indicated that virtually any human engineered organism is less fit than its natural counterpart in a natural environment' (Church, 2005). Let us go forth and safely multiply, *Nature*, 438, 24 November 2005, 423.

BfR. (2006). Press release of the German Bundesinstitut für Risikobewertung from 26 May 2006: Nano particles were not the cause of health problems triggered by sealing sprays! Available at http://www.bfr.bund.de/cd/7842

Brown, D. (2002). U.S. regulators want to know whether nanotech can pollute, Small Times, 8 March 2002. Available at http://www.electroiq.com/index/display/nanotech-article-display/267713/articles/small-times/environment/2002/03/bus-regulators-want-to-knowbrwhether-nanotech-can-pollute-b.html.

Chicken, J. and Posner, T. (1998). The Philosophy of Risk, Thomas Telford 1998, 7.

Drexler, E. (1986). Engines of Creation: The Coming Era of Nanotechnology, Anchor Press/Doubleday, New York.

Drexler, E. (1992). Nanosystems, Wiley Interscience 1992. Available at http://e-drexler.com/d/06/00/EOC/EOC_Table_of_Contents.html and http://e-drexler.com/d/06/00/Nanosystems/toc.html

Endy, D. (2005). Foundations for engineering biology, *Nature*, 438, 24 November 2005, 449–453.

ETC Group had issued a first communique as early as May 2002, titled 'No Small Matters!'

ETC Group. (2003). The Big Down. From Genomes to Atoms, January 2003. Available at http://www.etcgroup.org/upload/publication/171/01/thebigdown.pdf.

ETC Group. (2005). Nanotech's 'Second Nature' Patents: Implications for the Global South, May 2005. Available at http://www.etcgroup.org/en/node/54. A more moderate analysis is given in: Jones, R., Are natural resources a curse?, Nat. Nanotechnol., 2, November 2007, 665–666.

Freitas, R. (2000). Some Limits to Global Ecophagy by Biovorous Nanoreplicators, with Public Policy Recommendations, April 2000. Available at http://www.foresight.org/nano/Ecophagy.html

Garfinkel, M., Endy, D., Epstein, G., and Friedman, R. (2007). Synthetic Genomics: Options for Governance, J. Craig Venter Institute, CSIS and M.I.T. October 2007. Available at http://www.jcvi.org/cms/fileadmin/site/research/projects/synthetic-genomics-report/synthetic-genomics-report.pdf

GermanMedia. (1997–2000). Some examples from German media: Stern magazine from 18 February 1999, showing the space elevator with CNT cables on pages 100–101, as does the weekly *Die Zeit* from 23 November 2000 in a special dossier; both also featured nanomachinery images from Drexler's work. *PM* magazine showed medical nanobots in 1997, *Die Zeit* later even more extensively. German online magazine *Telepolis* showed in 1998 a microscopic submarine, referring to the 1966 film *Fantastic Voyage* and relating it explicitly to nanotechnology.

Joy, B. (2000). Why the future doesn't need us, Wired, April 2000. Available at http://www.wired.com/wired/archive/8.04/joy_pr.html

Kuhlbusch, T., Krug, H., and Nau, K. (ed.). (2009). NanoCare: Health Related Aspects of Nanomaterials. Final Scientific Report, July 2009. Available at http://www.nanopartikel.info/files/content/dana/Dokumente/NanoCare/Publikationen/NanoCare_Final_Report.pdf

Lux Research. (2007). Profiting from International Nanotechnology, March 2007.

Maurer, S., Lucas, K., and Terrell, S. (2006). From Understanding to Action: Community-Based Options for Improving Safety and Security in Synthetic Biology, White Paper, University of California at Berkeley, written as a policy draft for Synthetic Biology 1.0 conference. Available at http://gspp.berkeley.edu/iths/UC%20White%20Paper.pdf.

McKenna, P. (2008). Smelly sock treatment leaks silver nanoparticles, New Scientist, 7 April 2008. Available at http://www.newscientist.com/article/dn13602

Merrill Lynch. (2004). Merrill Lynch announced its *Nanotech Index* on 1 April 2004. Available at http://www.ml.com/index.asp?id= 7695_7696_8149_6261_13714_13728.

There were also the *Punk Ziegel Nanotechnology Index* and the *Lux Nanotech Index*.

NCI. (2003). This was a 'Challenge Goal' of the US National Cancer Institute (NCI), stated in 2003 by former NCI director Andrew von Eschenbach and later in http://nano.cancer.gov/objects/pdfs/Cancer_Nanotechnology_Plan-508.pdf

Nel, A., Mädler, L., Velegol, D., Xia, T., Hoek, E., Somasundaran, P., Klaessig, F., Castranova, V., and Thompson, M. (2009). Understanding biophysicochemical interactions at the nano-bio interface, *Nat. Mater.*, 8(7), July 2009, 543–557.

Nordmann, A. (2006). Personalisierte Medizin? – Zum Versprechen der Nanomedizintechnik, Hessisches Ärzteblatt, Mai 2006. Available at http://www.laekh.de/upload/Hess._Aerzteblatt/2006/2006_05/2006_05_08.pdf

NSTC. (2000). National Science and Technology Council, National Nanotechnology Initiative. The Initiative and Its Implementation Plan, July 2000, 19. Available at http://www.wtec.org/loyola/nano/IWGN. Implementation.Plan/nni.implementation.plan.pdf

Oberdörster, G., Oberdörster, E., and Oberdörster, J. (2005). Nanotoxicology: An emerging discipline evolving from studies of ultrafine particles, Environ. *Health Perspect.*, 113(7), July 2005, 823–839.

Paschen, H., Coenen, C., Fleischer, T., Grünwald, R., Oertel, D., and Revermann, C. (2003). TA-Projekt Nanotechnologie Endbericht, July 2003; German summary available at http://www.tab.fzk.de/de/projekt/zusammenfassung/ab92.htm

Poland, C., Duffin, R., Kinloch, I., Maynard, A., Wallace, W., Seaton, A., Stone, V., Brown, S., MacNee, W., and Donaldson, K. (2008). Carbon nanotubes introduced into theabdominal cavity ofmice show asbestoslike pathogenicity in a pilot study, *Nat. Nanotechnol.*, 3(7), July 2008, 423–428.

Roco, M. (ed.). (1999). Nanotechnology. Shaping the world atom by atom, National Science and Technology Council, December 1999. Available at http://www.wtec.org/loyola/nano/IWGN.Public.Brochure/IWGN. Nanotechnology.Brochure.pdf – for a critical review see Nordmann, A., Shaping the World Atom by Atom: Eine nanowissenschafliche Bildwelt-analyse, November 2002, in: Grunwald, Armin (ed.)., Technikgestaltung zwischen Wunsch und Wirklichkeit, Springer, 2003, 191–199.

Roco, M. and Bainbridge, W. (ed.). (2001). Societal Implications of Nanoscience and Nanotechnology, National Science Foundation, March 2001, 3.

Royal Society and Royal Academy of Engineering, Nanoscience and Nanotechnologies. (2004). Opportunities and Uncertainties, July 2004. Available at http://www.nanotec.org.uk/finalReport.htm – recommendations are listed in chapter 10, pp. 85–87.

SA. (2001). The quoted question appeared on the cover of *Scientific American*, September 2001, which was a special issue dedicated to nanotechnology.

Schmundt, H., and Verbeet, M. (2008). Kleine Teilchen, großes Risiko, Der Spiegel 24/2008, 148–150; online.

Schummer, J. (2009). Nanotechnologie. Spiele mit Grenzen, edition unsold 2009, 58.

Taniguchi, N. (1974). On the basic concept of 'nanotechnology', Proc. Intl. Conf. Prod. Eng. Tokyo, Part II, *Jpn. Soc. Prec. Eng.*, 1974.

TE. (2006). Has all the magic gone?, *The Economist*, 8/15 April 2006. Available at http://www.innovationsgesellschaft.ch/images/fremde_publikationen/HasalltheMagicgone.pdf

Toxicology issues of nanoparticles in general were discussed in Rotman, D., Measuring the risks of nanotechnology, *Technol. Rev.*, April 2003.

Venter, C. (2008). Interview for *Newsweek*: Zakaria, F., A Bug to Save the Planet, *Newsweek*, 16 June 2008. Available at http://www.newsweek.com/id/140066

Wick, P., Manser, P., Limbach, L., Dettlaff-Weglikowska, U., Krumeich, F., Roth, S., Stark, W., and Bruinink, A. (2007). The degree and kind of agglomeration affect carbon nanotube cytotoxicity, *Toxicol. Lett.*, 168, 30 January 2007, 121–131.

Woodrow Wilson Center. (2010). Nanotechnology Consumer Products Inventory. Available at http://www.nanotechproject.org/inventories/consumer/

Wolfe, J., Paull, R., Hébert, P., Sinkula, M., Mittal, N., Seltzer, R., Gosalia, D., Greenbaum, J., Kamen, E., and Stallone, M. (2003). The Nanotech Report 2003, Lux Capital, 2003; it states: 'Mentions of 'nanotechnology' in the popular press have surged 2,000% from approximately 200 in 1995 to more than 4,000 in 2002.'

2.5 Public Perception of Nanotechnologies: Challenges and Recommendations for Communication Strategies and Dialogue Concepts

Dr. Antje Grobe and Nico Kreinberger
University of Stuttgart, Germany

2.5.1 *Introduction*

Nanotechnologies are estimated as being one of the key drivers for innovative technologies for the sustainable reduction of energy consumption (BMBF, 2009). This includes a broad field of applications for energy supply and energy storage, such as photovoltaic technologies, fuel cells, lightweight and high-strength materials for rotor blades of windmills, technologies for lithium-ion batteries, and hydrogen storage – just to name some of them. Different nanomaterials are applied as coatings for easy-to-clean surfaces, which could help to reduce the consumption of water, energy, and chemicals; as construction materials or, more generally, for example, in the case of natural polymers; and as new CO_2-neutral materials, which could replace chemical products based on oil. To look at only one of these examples would mean to open a specific microcosmos of possible materials, technologies, and ways of processing or applications. This diversity and complexity is typical for nanotechnologies (Renn and Grobe, 2010). Maybe, this is one of the reasons why a description of nanotechnologies seems always to be incomplete and why it is difficult to have a clear picture of what 'nanotechnologies' really mean. Even for public perception it is a crucial issue if they can find appropriate mechanisms to cover this complexity.

The following sections will try to have a look behind the typical psychological, social, and cultural factors which are basic for public perception of nanotechnologies in general (see Section 2.5.2).

The perception of applications in a specific area such as energy is embedded in the context of general expectations and attitudes towards nanotechnologies (see Section 2.5.3) and a certain knowledge base about the different applications (see Section 2.5.4). It will be explained what kind of representations (images, examples, and stories) consumers are connecting in their minds with the term 'nanotechnologies' – even with the special focus of energy (see Section 2.5.5). Besides the question, what consumers know about this issue, we will ask, what consumers want to know (see Section 2.5.6) and what kind of recommendations can be given for consumer communication and dialogue concepts (see Section 2.5.7). The following chapter is based on the study 'Nanotechnologies: What Consumers Want to Know,' which was published only in German language on behalf of the Federation of German Consumer Organisations in 2008 (Grobe *et al.*, 2008).

2.5.2 *Psychological, Social, and Cultural Factors of Risk Perception*

The first and crucial question if you are talking about the perception of nanotechnologies is, are they perceived as a benefit or as a risk? Psychology and social sciences described *risk perception* as a process in which a physical signal or information about potential hazardous events or activities is getting recognised. This process includes an assessment of the estimated seriousness, range of outcome, probability of occurrence, the specific concernment, and its acceptability (Slovic, 1992). Every factor, which is relevant for the assessment of technologies, consists of a whole set of criteria that can have influence on the results. Such factors depend on the individual skills regarding the attention and perception which were caused by a specific stimulus or theme, individual knowledge, attitude and experience, feelings, social shared cultural values, pictures, and stories. The assessment is based on the comparison of individual and common interpretations, tightened during a permanent process of cultural learning, modification, or rejection (Covello, 1983; Jungermann and Slovic, 1993; Renn, 2004; Renn, 2005; Sjöberg, 2006; Slovic, 1992).

If you compare the different psychological, sociological, and cultural studies, it is possible to compress several factors of influence from the above-mentioned literature:

- Societal background attitudes such as risk aversion, risk acceptance, or neutrality
- Personal and societal risk-benefit expectation
- Rating of the degree of uncertainty and the lack of knowledge
- Presumption of exposition by a hazard source
- Probability of occurrence, estimated number, and degree of damage
- Possibility of compensation
- Degree of trust and customisation in handling a risk source
- Possibility of personal control/control by other institutions and actors
- Trust in institutions and actors
- Voluntariness/freedom of choice
- Imbalance in disposition of risks and benefits
- Artificiality of risk source/naturalness
- Accountability of a default
- Emotional quality of a theme
- Possible stigmatisation of debate

This summary of psychological and sociological evaluation criteria for the consideration of benefits and risks shows: 1) Simple solutions for how the awareness of benefits and risks could be influenced, rather in a positive or in a negative way, do not exist; and 2) the key to personal and societal shared evaluations could be the *accessibility of information*. Without access to balanced information, it is neither possible to reduce the uncertainty of a benefits-and-risk ratio or to estimate the lack of knowledge nor possible to generate trust in the actors and their handling of risks. All expectations and presumptions are dependent on societal shared information about the benefits and risks and about the actors from science, industry, public authorities, and critical non-governmental organisations

(NGOs) such as consumer organisations, environmental groups, trade unions, and churches. Voluntariness, or better, the customers' freedom of choice, is not possible without information if the product they bought is a 'nanoproduct', about what materials or technologies are used, what kind of properties they have, and if they are safe.

Trust and a feeling of possible control are directly addicted to information about the actors, too.

That causes an interesting interdependency: the perception that *information is rare evokes a trend to expect more risks*, due to the fact of uncertainty (Rubik and Weskamp, 1996). *The higher the expected risk* is estimated and the feeling of personal uncertainty increases, *the higher the need to get more information*. Unfortunately, at the same time the recognised *need for more information* is coupled with the suspicions that the information you can gain could be *insufficient.* This is first and foremost the case, when there is uncertainty if the source of information is trustworthy. In that case *information will always be insufficient* as long as the tendency for mistrust is still dominant.

In principle, this situation is avoidable and not irreversible. Researchers and companies can communicate proactively about what they are doing and could give more information to ensure the safety of their products or intended outcomes of the research. Easy accessibility of information and open communication from different stakeholders about the quality of information could help not to accelerate a downward spiral of increasing uncertainty.

So far this is the theoretical approach. However, one core question remains critical: do consumers believe the sender of the information? And, what is the societal background of each of the interpretations of information which could balance the feeling of uncertainty? The following sections will ask about the general public perception of nanotechnologies in an international comparison and clarify if this theoretical vicious circle has just started.

2.5.3 *Public Perception of Nanotechnologies: an International Comparison*

Nanotechnologies are promised to be one of the key technologies of the 21st century (BMBF, 2006; BMBF, 2009). However, diverse

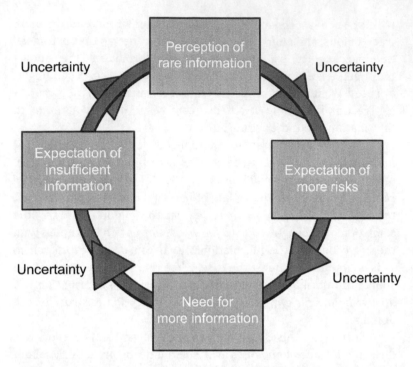

Figure 2.5 Vicious circle of insufficient information. See also Colour Insert.

national and international representative consumer surveys showed that there is still a lack of knowledge at the consumer's side. In 2008 the International Risk Governance Council (IRGC) published a study which compares international studies about the perception of nanotechnologies (IRGC, 2008). In 2004, only 16% of the American respondents of a random sample telephone survey had heard 'some' or 'a lot' about nanotechnologies (Cobb and Macoubrie, 2004). For the United Kingdom, a study on behalf of The Royal Society and the Royal Academy of Engineering summarised that only 29% of the British public was aware of nanotechnologies in 2004 (The Royal Society, 2004). The first German survey occurred in the same year with a nearly balanced result. Surprisingly, 45% knew about this technology and could give specific information about it. Only 48% had never heard of nanotechnologies (komm.passion, 2004). In the United States, a similar increase of popularity up to 40% was measured in 2005 (Einsiedel, 2005; Macoubrie, 2005). In the same year, 38% of the Canadians could give more detailed information regarding nanotechnologies (Einsiedel, 2005). In September 2006, another representative survey was carried out (Hart, 2006). Public awareness and the degree of information went down to 30% who knew 'some' or 'a lot' and 69% had heard little or nothing about nanotechnologies. This tendency of a decreasing level of awareness and knowledge was adjusted in a Woodrow Wilson report in 2007 (Kahan *et al.*, 2007). Just around 20% of the respondents indicated to know 'some' or 'a lot' and could specify the knowledge by applications or maybe by a definition (Figure 2.6).

A second representative survey with 1,000 persons was realised by the German Federal Institute for Risk Assessment in 2007 and published in 2008 (BfR, 2008). Fifty-two percent of the public declared to know 'some' or 'a lot' about nanotechnologies and could give specific examples. With 48% of the public that said it knew little or nothing, Germany had the most positive public perception in this comparison of international data. Nevertheless, there are still half of the people who have no or insufficient information about nanotechnologies – and the data does not suggest that we are in an upward trend. The latest survey from the United States (Hart, 2009) showed again similar data to the 2006 polls. About 68% had heard a little or nothing, and 31% claimed to know 'some' or 'a lot'. Most of the surveys asked the consumers – besides general awareness – about their attitudes towards nanotechnologies. The core question was if no or less information about this subject would be connected with negative attitudes or mistrust or if the positive expectations would outweigh the risks. A closer view of the expressed expectations shows a diverse picture.

Figure 2.7 shows a broad range and volatility of results. One possible reason might be the different methodological designs that were used in the surveys. Some asked only for expected benefits in relation to risks; others asked also for ambivalent expectations, or they had a 'don't know' option. In the latest US survey (2009) form 'Hart Research Associates', no information is given about the perceived benefits or risks. With this database, general assumptions have to be taken with care, but it seems that the consumer's attitudes are still not established on this issue. Interestingly, with the turn to

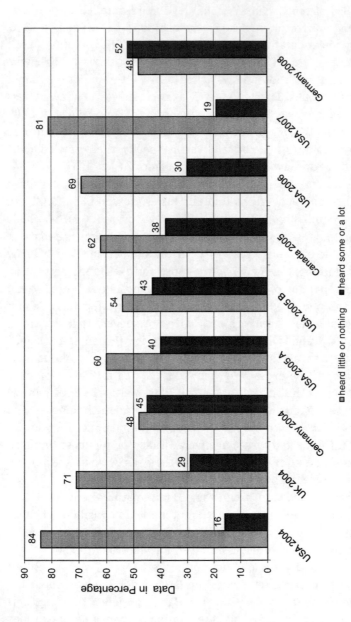

Figure 2.6 Public knowledge base on nanotechnologies in international surveys. See also Colour Insert.

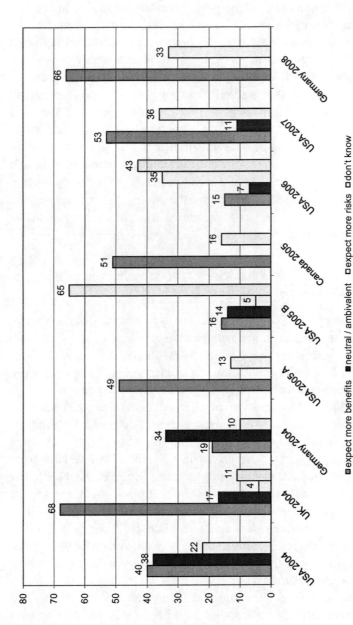

Figure 2.7 Public expectations on benefits and risks. See also Colour Insert.

a possible increase of more negative attitudes, the engagement of funders to investigate public perception turned down. In the last few years, no representative surveys have been published in the countries which were active in the years before – besides the United States. Therefore, the question how attitudes recently developed, especially in the UK, Germany, and Canada, has to remain open.

The situation that at least half of the public is not at all or insufficiently informed about nanotechnologies, and that it is nearly impossible to have a clear picture about its preferences and attitudes, should be a serious indicator that too less attention has been drawn to the issue of information, communication, and dialogue. Coming back to the above-mentioned chance to avoid a vicious circle of insufficient information, rising risk perception, and increasing mistrust, it has to be stated that the efforts should be strengthened as soon as possible.

2.5.4 *Consumer's Perception of Nanotechnologies in German Language Areas*

For Germany, the comparison of surveys showed a quite positive result. The next sections will provide a view, in depth, of the consumer's attitudes towards nanotechnologies in general and later on towards specific applications in connection with energy supply, storage, or the reduction of energy consumption in general.

In 2008, the Federation of German Consumer Organisations (VZBV) initiated a study on consumer perceptions of nanotechnologies. One hundred qualitative in-depth interviews with consumers in Germany and the German-speaking area of Switzerland were conducted by the University of Stuttgart (Grobe *et al.*, 2008). The aim of the qualitative survey was to explore what consumers know about nanotechnologies, what they want to know, how they want to be informed, and who should deliver what kind of information. Therefore, only people who were aware about the term 'nanotechnology' participated – independent from the question how much they knew about it.

In the beginning of the interviews, consumers were asked about their imaginations or ideas about nanotechnologies. Starting with this open question, researchers followed the examples, issues, comparisons, and qualitative assessments of different nanotechnology applications which were mentioned by the respondents themselves. Researchers just deepened the suggested themes. With this open and narrative method, the societal framing, reference systems, and priorities of consumers could be identified.

Interestingly, low significance was observed concerning the differences in the attitudes or knowledge base of the well-balanced sample (age, gender, education): Neither the level of education nor the age or cultural differences between German and Swiss participants exert a considerable influence on the general knowledge or the attitudes. The researchers explained this observation by the preferred ways of information. As the most frequently used information medium, science shows on TV were mentioned. Theses science shows were perceived very well by a broad group of participants. Here, the influence of German television formats was visible in the Swiss sample (German-speaking area) in nearly the same way. Gender triggered to a low extent the number of known applications. Men knew rather eight applications; women had an average of six entries. Only in the question of self-estimation of the own knowledge base did a difference appear – only men said that they really know a lot about nanotechnologies (Grobe *et al.*, 2008: 19) (see Figure 2.8).

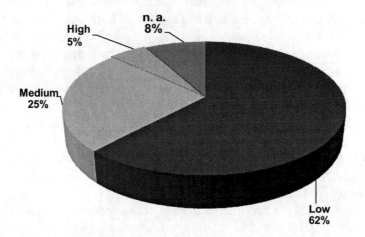

Figure 2.8 Consumer's self-estimation of knowledge about nanotechnologies (Grobe *et al.*, 2008). See also Colour Insert.

But, besides this detail, 62% of the consumers assessed their own knowledge base as rather 'low'. Most of the interviews just started with a quote such as, 'I don't know a lot about it, but …', and then started with descriptions of different applications.

The same consumers, who declared to have a low knowledge base, explained in the following interviews a large number of different applications. The average was around seven to eight fields of application each of the interviewees mentioned (Figure 2.9).

About 85% of the respondents mentioned *medical applications*. Many informed themselves actively about cancer treatments, catheters and stents, or implant technologies due to personal concerns in their families. Several articles in the print media or reports in the news raised a lot of attention coupled with mainly positive expectation.

Surface coatings were highlighted by 78 persons and co-notated positively by the large majority. Main associations based on the lotus blossom and its ability to clean itself. Consumer found a lot of arguments, such as the reduction of chemicals or cleansing agents, and they underlined their positive expectation towards this kind of convenient applications of nanotechnologies.

The area of *food* was mentioned by 63% of the respondents. People picked up a lot of ideas for possible usages such as colour- or taste-changing beverages or food, artificial meat from a kind of 'nanowave', or chocolate coatings. Some others mentioned the positive effects of a better convenient nutrition. Most of the examples alternate between fantasy and reality. Additionally, it has to be mentioned that food holds most of the negative comments in comparison to the other mentioned applications. Fears of negative health effects and a clearly expressed mistrust against the food industry were documented as reasons.

In the *automotive industry* the scope of interest was related to scratch-resistant and easy-to-clean coatings or cleansing products which were tested by several of the respondents. Only 1 out of 62 interviewees mentioned lightweight materials and the reduction of energy consumption as a benefit.

Information and communication technologies and electronics were considered by 61%. Here, nanotechnologies were seen as

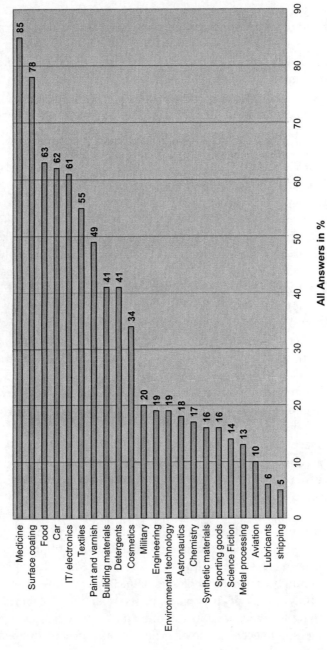

Figure 2.9 Mentioned areas of nanotechnology applications.

another step in the process of miniaturisation and maximisation of efficiency.

These 'top 5' of the most mentioned applications of nanotechnologies were followed be functional textiles, easy-to-clean or anti-fouling facade paints, construction materials (windows, concrete, thermal insulation materials), detergents, cosmetics, military uses, engineering, environmental technologies, space technology, chemicals and plastics, sporting goods, science fiction, metal processing, aviation, lubricants, and anti-fouling coatings for the shipping industry.

All in all, the German and Swiss consumers were surprisingly well informed. Nanotechnologies were examined, differentiated in dependency on a certain application. Most of the negative comments went to the field of applications in the food industry (16%), followed by military uses (12%). Other participants talked about risks, for example, in connection to pharmaceuticals, detergents, or cosmetics. Those remarks occurred by 2 up to 4% of the interviewees. But compared with the positive appraisals for surface coatings (51%), medical applications (46%), the automotive sector (39%), paints (33%), and construction materials (27%), the negative comments had a rather minor significance. In this survey from 2008, consumers co-notated concrete applications – aside from the above-mentioned examples – in a more positive way.

If consumers were asked to summarise their attitudes towards nanotechnologies, nearly two-thirds of the respondents expressed a positive point of view (see Figure 2.10).

However, the same group of respondents *did refer to risk* with a surprising high number of quotes – but these remarks were more general and by the majority not directly connected to certain applications.

About 87% were generally afraid of negative health effects, and 29% mentioned possible harms of the environment caused by nanotechnologies. A 'Damocles sword' of (the military's or a terrorist's) misuse was indicated by 11%, and 6% were concerned by the ubiquity of nanotechnologies in the large number of everyday items. Only 2 out of 100 participants referred to cyborgs and converging technologies. On the one hand it was interesting that consumers connected positive expectations, such as 'innovation',

Would you try a nanoproduct? N=100

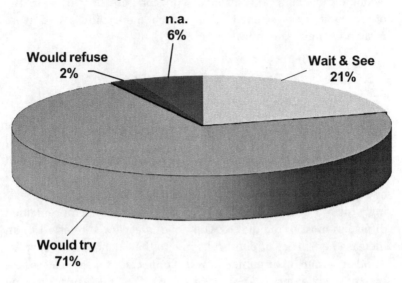

Attitudes towards nanotechnologies? N=100

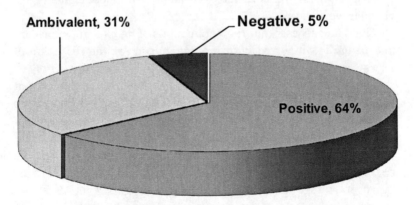

Figure 2.10 General consumer attitudes towards nanotechnologies. See also Colour Insert.

'convenience', 'healthy and beauty', a 'positive environment' or 'economic benefits', and 'safety', with concrete products. On the other hand, risks seemed to be much more diffuse and were mentioned more generally.

2.5.5 Attitudes Towards Nanotechnologies in the Energy Sector

Most of the quotes which indicated energy consumption or the reduction of resources were connected with applications in the field of *environmental technologies* out of 19 respondents who mentioned environmental technologies expected an increase of renewable energies, such as photovoltaics, through nanotechnologies. Interestingly, the role of nanotechnologies in photovoltaic systems remained diffuse in most of the quotes. Only one interviewee referred to an increased efficiency of photocells due to their nanosurface coating. Another group of arguments was connected to applications in the *construction area*. As written earlier, thermal insulation, new windows, and new materials or technologies for low-energy houses were discussed by the consumers. In the *automotive sector*, new traction technologies and low-friction bearings were explained by two participants – but both had experiences with these technologies via their workplaces.

The data suggests that the applications of nanotechnologies are not primarily connected with arguments such as the reduction of energy consumption or with new technologies for energy supply or storage. In other words, the main arguments of sustainability are not well represented in the communication patterns of the respondents in the 2008 survey.

Bonazzi listed in his publication 'Communicating Nanotechnology', edited by the European Commission, several sustainable energy applications of nanotechnologies: solar energy; fuel cells; hydrogen storage and hydrogen production; lightweight materials, which reduce the energy consumption in mechanical processes; insulation technologies, and a more sustainable consumption of resources due to different surface technologies (Bonazzi, 2010: 36). The author identified 'nano-energy' and 'nano-environment' as the second of three 'main areas for urgent communication to selected audiences' (Bonazzi, 2010: 35). They connected nano-energy and environmental technologies clearly to sustainability as 'one of the issues of major public and policy concern'.

It is a good question why policy makers, the industry, NGOs, or the media passed to communicate about this kind of applications and its benefits. In 2008, the term 'sustainability' was not mentioned by the consumers in the German qualitative survey at all. However, consumers told the researchers a lot about what they would like to be informed about and how. The following section will deepen these questions in general and will develop several recommendations which are suitable for the area of energy applications, too.

2.5.6 *Requirements for Consumer Communication*

In the Hart's US survey about nanotechnologies, 90% of the respondents of a telephone poll answered that 'more should be done to inform the public about this research' (Hart, 2006: 16). The German qualitative survey (Grobe *et al.*, 2008) collected a lot of quotes from the consumers about whom, how, and about what they would like to be informed about nanotechnologies. Science (53%), industry (51%), public authorities (35%), and consumer organisations (31%) were identified as desirable sources of information and had been requested to prepare better, easy-to-access, and easy-to-understand information. Twenty-eight percent claimed the media should inform them. Personal networks and environmental organisations were stated by 20% of the consumers as their favourite source of information.

The data and the qualitative quotes suggest that participants tend to choose more than one source of information. They call for more information by the industry about their products and for information from science about risk appraisal or new research results. Public authorities (35%) and consumer organisations (31%) were seen as other important actors for the consumers. Twenty-five percent expressed a high trust in science, followed by 22% for consumer organisations, public authorities, and the industry (both 18%). Environmental organisations were often perceived as campaigning groups. Fifteen percent of the interviewees expressed their trust towards this information source. Beside the question of

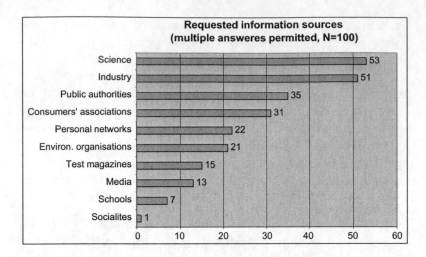

Figure 2.11   Requested sources of information.

the sender, a lot of respondents elaborate on the way they want to be informed (see Figure 2.11).

In the first instance, people highlighted the important role of television as their favourite information channel. A lot of the participants reported from science shows and news formats which were very attractive for them. Animations of basic functionalities or scientific principles had been indicated as valuable. Traditional print media (46%) were followed by the internet (40%). Here, independent information platforms were requested – and of course, easy-to-access information about concrete products. Thirty-eight percent said that products created by nanotechnologies and nanomaterials should be labelled, and 28% asked for specific information about the safe application, care, and disposal of nanoproducts on the back of the packaging. Eighteen percent wanted to be better informed by advertising.

Via these channels consumers want to know something about the functionality, properties, and effects of nanomaterials and nanotechnologies in the concrete product (57%). Information about the safety of the products and materials were indicated by 47% of the respondents. In the view of 36%, the ingredients (e.g., in food and cosmetics) should be named out – and of course the

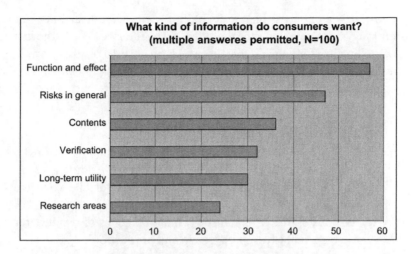

Figure 2.12   Kind of requested information.

information if a product contains nanomaterials at all. Thirty-two percent strengthened that it would be not sufficient if the producers would claim their products as being safe without an independent approval of this statement. Additionally, information should be given about the long-term effects (30%) – again adjusted by a neutral organisation. At least, people were interested to know more about future developments and fields of research (24%) (see Figure 2.12).

These results could encourage scientists, the industry, public authorities, NGOs, and the media to deepen the information about nanotechnologies in a more detailed way. Consumers expressed a serious need for those details. The findings about the number of applications and the degree of differentiation exhibit that consumers are well prepared to handle the complexity of nanotechnologies and that they remember this information in a balanced way. For the above-mentioned actors these issues should be the core challenges for their communication strategies.

2.5.7   *Conclusions: Recommendations for Communication Strategies and Dialogue Concepts*

The majority of consumers expressed in the international surveys that the knowledge about nanotechnologies is still insufficient as two-thirds – or, in the best case, half of them – had heard little or nothing about it. No wonder that 90% of the interviewees in the last US polls called for better information. As the German results showed, even those who had heard something about nanotechnologies estimated their own knowledge base as rather low and asked for more information about:

- Function and properties (long term)
- Benefits (including a quantification of sustainability issues)
- Ingredients/substances/technologies (nano inside)
- Risk assessment and safety (carried out for the concrete product, independently controlled)
- Responsibility (personalised, taking up codes of conduct for the responsible research, usage and production)

To prepare these different types of information, actors would be well advised to take part or to invest in stakeholder and public dialogue formats. Expectations and needs vary – as far as the data from Germany and Switzerland suggests – from application to application; they differ between the stakeholders and sometimes even from actor to actor.

Additionally, cultural differences have to be respected. Maybe, industry and consumer associations could support their members to invest in sector-specific dialogue exercises with the consumers to elaborate their specific needs for information and appropriate channels such as TV, print media, or the Internet. If this will not happen in the near future, it will be even harder to escape from the downward spiral of uncertainty, the remaining feelings of insufficient information, and mistrust. Applications of sustainable, sense-making research projects, and products in the area of energy supply and storage of renewable energies could provide the needed examples to escape from this tendency which has just started.

Bibliography

Bonazzi, M., (2010). Communicating Nanotechnology. Why, to Whom, Saying What and How? Edited by the European Commission DG RTD, Directorate Industrial Technologies. Brussels.

BfR. (2008). Federal Institute for Risk Assessment, Edited by René Zimmer, Rolf Hertel, Gaby-Fleur Böl, Public Perceptions about Nanotechnology, representative survey and basic morphological-psychological study, Berlin 2008 (BfR-Wissenschaft 01/2009). Available at http://www.bfr. bund.de/cm/290/public_perceptions_about_nanotechnology.pdf

BMBF. (2006). Federal Ministry of Education and Research: Nano-Initiative – Aktionsplan 2010, Berlin.

BMBF. (2009). Federal Ministry of Education and Research: Nano.DE-Report 2009, Status Quo of Nanotechnology in Germany, Bonn.

Cobb, M. D. and Macoubrie, J. (2004). Public perceptions about nanotechnology: risk, benefits and trust, *J. Nanopart. Res.*, 6, 395–405.

Covello, V. T. (1983). The perception of technological risks: a literature review, *Technol. Forecast. Social Change*, 23, 285–297.

Einsiedel, E. (2005). In the public eye: the early landscape of nanotechnologies among Canadian and U.S. publics, Online J. Nanotechnol., 1.

Grobe, A., Nawrath, S., Rekic, M., Schetula, V., and Schneider, C. (2008). Nanotechnologien. Was Verbraucher wissen wollen. Studie im Auftrag des Verbraucherzentrale Bundesverbandes e.V., edited by VZBV, Berlin.

Hart, P. D. (2006). Hart Research Associates, Report Findings, Conducted on behalf of the Woodrow Wilson International Center for Scholars, Project on Emerging Nanotechnologies, New York, Woodrow Wilson International Center for Scholars.

IRGC. (2008). Appropriate Risk Governance Strategies for Nanotechnology Applications in Food and Cosmetics. Genf: International Risk Governance Council, written by Antje Grobe, Ortwin Renn and Alexander Jaeger.

Jungermann, H. and Slovic, P. (1993). Charakteristika individueller Risikowahrnehmung. Edited by W. Krohn and G. Krücken: Riskante Technologien: Reflexion und Regulation.

Kahan, D. M., Slovic, P., Braman, D., Gastil, J., and Cohen, G. (2007). Nanotechnology Risk Perceptions: The Influence of Affect an Values, in: Woodrow Wilson International Center for Scholars.

Komm.passion. (2005). Wissen und Einstellungen zur Nanotechnologie, Frankfurt.

Macoubrie, J. (2005). Informed Public Perceptions of Nanotechnology and Trust in Government, Edited by Woodrow Wilson International Center for Scholars. Washington, D.C.

Renn, O. (2004). Perception of risks, *Geneva Papers Risk Insurance*, 29(1), 102–114.

Renn, O. (2005). Risk perception and communication: lessons for the food and food packaging industry, in: *Food Additives Contaminants*, 22(10), 1061–1071.

Renn, O. and Grobe, A. (2010). Risk governance in the field of nanotechnologies: core challenges of an integrative approach, in: International Handbook of Regulating Nanotechnologies, edited by Graeme A. Hodge, Diana M. Bowman and Andrew D. Maynard, Edward Elgar Cheltenham, UK, Northampton, MA, USA, 484–507.

Royal Society. (2004). Royal Society and the Royal Academy of Engineering Nanotechnology Working Group, Nanotechnology: Views of the General Public, Quantitative and Qualitative Research, carried out as a part of the Nanotechnology Study, London, Royal Society.

Sjöberg, L. (2006). Rational risk perception: Utopia or dystopia? *Risk Res.*, 9(6), 683–696.

Slovic, P. (1992). Perception of risk: reflections on the psychometric paradigm, in: S. Krimsky and D. Golding (eds.). Social Theories of Risk, Westport, Praeger, 117–152.

Slovic, P., Fischhoff, B., and Lichtenstein, S. (1980). Facts and fears: understanding perceived risk, in: R. Schwing and W. A. Albers (eds.). Societal Risk Assessment: How Safe is Safe Enough? New York: Plenum, 181–214.

Woodrow Wilson International Center. (2009). Nanotechnology, Synthetic Biology, & Public Opinion, a report of the findings, based on a national survey among adults, conducted on behalf of Project on Emerging Nanotechnologies, the Woodrow Wilson International Center for Scholars by Hart Research Associates. Washington, D.C.

Chapter 3

Examples for Nanotechnological Applications in the Energy Sector

Nanotechnology can be applied to the whole value chain of the energy sector: energy sources – energy conversion – energy distribution – energy storage and energy use. The overview of the possible application range of nanotechnology in the energy sector (see Figure 1.7) shows the vast and broad variety of possible products and applications. In this chapter selected examples of nanotechnological applications and innovations are described in detail. It starts with an overview about the possible use of aerogels. The variety of nanotechnology can be particularly exemplified by aerogels. In the field of energy sources and conversion, the current state of technology and market development of the dye solar cell (DSC) technology are presented. Another section deals with the possibility to recover waste heat by the use of nanoscale thermoelectrics. Furthermore nanostructured ceramic membranes are described because membrane technology is discussed to play an important role in future carbon capture and storage (CCS) concepts leading to low CO_2 emission power generation. Nanomaterials for energy storage are a matter of another section, followed by three articles in the field of energy use. The first describes nanotechological

Nanotechnology and Energy: Science, Promises, and Limits
Jochen Lambauer, Ulrich Fahl, and Alfred Voß
Copyright © 2013 Pan Stanford Publishing Pte. Ltd.
ISBN 978-981-4310-81-9 (Hardcover), 978-981-4364-06-5 (eBook)
www.panstanford.com

applications and innovations in construction. The second shows how nanotechnology can be applied to manufacture active windows for daylight-guiding applications. The third focuses on energy efficiency potentials of nanotechnology in production processes.

3.1 Aerogels: Porous Sol-Gel-Derived Solids for Applications in Energy Technologies

Dr. Gudrun Reichenauer
ZAE Bayern, Germany

Many applications profit from the recent developments in the field of nanosized particles and nanostructured materials (Baxter *et al.*, 2009). Similarly, materials with high porosity in the nanometer range can be applied to specifically enhance certain effects and, thus, to increase the efficiency in different processes along the chain of providing, managing, and efficiently using energy. One class of nanoporous solids are sol-gel-derived porous materials, also known as *aerogels*.

3.1.1 *Aerogels–Synthesis and Properties*

3.1.1.1 Synthesis

The synthesis of aerogels involves two major steps, the sol-gel process and the drying of the gel (Brinker and Scherer, 1990; Hüsing and Schubert, 2005; Reichenauer, 2008). In the first step nanosized sol particles are formed, which interconnect upon the sol-gel transition to a three-dimensional (3D) open-porous backbone (Figure 3.1). All structural characteristics of the resulting gel, such as the size and shape of the particles, the size of the pores filled with the solvent, the solid to solvent ratio, and the linking of the particles, are controlled by the reactants and the process conditions in the sol-gel process (see e.g., Brinker and Scherer, 1990; Brinker *et al.*, 1994). Sol-gel processing can be applied to synthesise inorganic (metal oxide–based) as well as organic gel backbones (Figure 3.2).

By applying an additional pyrolysis step, organic precursors can be converted into synthetic porous carbon with an almost identical

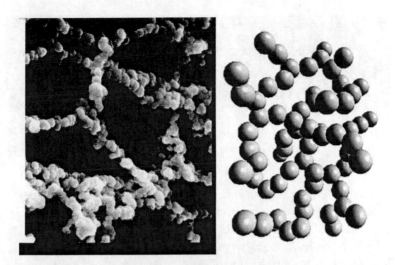

Figure 3.1 TEM image (left) and schematic model (right) of an aerogel consisting of particles interconnected to a three-dimensionally linked open porous backbone.

Figure 3.2 Left: Silica aerogel granules; right: organic, phenol-formaldehyde-based aerogel cylinder cut in half (© ZAE Bayern). See also Colour Insert.

meso- and macrostructure,[1] however additional microporosity (pores < 2 nm). Since the process is starting from a solution, it

---

[1] According to the International Union of Pure and Applied Chemistry (IUPAC), micropores are defined as pores < 2 nm, mesopores are characterized by pore sizes between 2 and 50 nm, and macropores are pores > 50 nm.

Figure 3.3 Left and middle: Organic aerogels with different types of fibres incorporated; right: hybrid aerogels consisting of an organic and a silica phase (© ZAE Bayern).

allows to easily introduce additives such as functional (nano) particles, fibres, or precursors for metal components (Figure 3.3). Infiltration of porous matrices such as foams, felts, or thin layers of fabric or paper with the starting solution is another option. A big advantage is also the shaping of the gels via moulding.

The second and very crucial process step is the drying of the gel (Rigacci *et al.*, 2003). In general the goal is to preserve the backbone structure upon drying and, thus, the pore network present in the gel. The main problem hereby is the low stiffness of the sparse gel backbone in combination with nanosized pores.

Upon convective drying, capillary pressures of up to 10 MPa are impounded onto the delicate backbone, thus resulting in a large compression of the gel. This results in reactive groups at the inner surface getting close to each other and forming irreversible links. As a consequence, most of the initially designed porosity is lost.

An alternative is the chemical modification of the inner surface, thus replacing highly reactive by non-reactive groups. This does not prevent the large compression of the gel upon convective drying; however, when the capillary forces are decreasing towards the end of the drying step, the gel re-expands (Schwertfeger, 1991; Prakash *et al.*, 1995) to its initial size and shape just like a flexible organic foam.

Other options for drying are freeze drying (Kocklenberg *et al.*, 1998; Tamon *et al.*, 1999; Babic *et al.*, 2004), which, however, mostly introduces significant structural changes, and the use of aging steps (Davis *et al.*, 1992; Haereid *et al.*, 1995; Wiener *et al.*, 2003; Reichenauer, 2004; Rigacci *et al.*, 2004) or additives that increase the stiffness of the gel backbone, thus preventing significant compression.

The classical approach to overcome this effect is supercritical drying (Kistler, 1932; Poco *et al.*, 1996) of the gels; this way small to negligible shrinkage upon drying can be achieved and the solid backbone is almost unaffected by the drying step.[2]

3.1.1.2 Structural properties

Aerogels are highly porous solids with a disordered[3] structured backbone defining a well-accessible and interconnected porosity. In contrast to well-ordered, for example, templated powders, aerogels can easily be provided as macroscopic components without the need for an additional binder phase or matrix. Due to the underlying synthesis route, porous solids with high chemical purity can be prepared if necessary. The porosity and the specific surface area of aerogels can be adjusted independently over a wide range, thus allowing control of the average pore and backbone particle size.

The pores in aerogels can be divided into micropores, mesopores, and macropores. One big advantage compared with conventional foams is the fact that aerogels allow providing mesoporous materials. On the other hand, when comparing aerogels with classical ceramics, far larger porosity at the mesopore scale can be achieved. If required the porosity of aerogels can also be reduced in a defined manner if the surface area per volume is to be optimised.

Typical values for mesoporosity range from 60 to 95%; however, even higher, up to 99%, or lower porosities can be provided as well. Specific surface areas (not taking into account surface area in micropores) can be adjusted from a few square meters per gram (in particular in organic and carbon aerogels) to about 700 m^2/g. In inorganic systems pores sizes are mainly in the mesopore to lower macropore range unless hierarchical structures are initiated (Brandhuber *et al.*, 2005); in contrast, in organic and carbon aerogels average pore sizes between a few nanometers only and 100 microns can be achieved.

---

[2] Originally all porous materials derived by supercritically drying were denoted as *aerogels*, while convectively dried gels were called *xerogels*. Nowadays most porous, sol-gel-derived solids are denoted as aerogels.

[3] Recently, aerogels with a highly ordered structure within their backbone were successfully synthesized (e.g., Brandhuber *et al.*, 2005).

Isotropic or anisotropic functionalities can be introduced by additional components in the form of additives, matrices, or coatings. Hereby components can be integrated by dispersion of additives in the starting solution, by infiltration of the gel or the aerogel via the liquid or the gas phase, or by bottom-up processes where the additional component is formed *in situ* from molecular precursors added to the starting solution.

3.1.2 *Properties Meeting Applications*

The specific structural properties of aerogels can be subdivided into different application related classes.

High porosities combined with pores smaller than about 100 nm are required for thermal insulation, while large, well-accessible surface areas are the prerequisite for catalyst supports as well as charge and adsorption storage devices. Other potential applications in energy technology are filters or gas separation components. Furthermore the fact that porosities, pore sizes, and surface areas can be tailored over a wide range also make aerogels excellent model systems that allow for the identification of mechanisms that can then be applied to optimise other porous systems.

3.1.2.1 Thermal insulation

Thermal transport in an aerogel can be described by the super-position of three heat transfer mechanisms (Fricke, 1994; Ebert, 2011), that is, transport along the solid phase, heat transfer via radiation, and thermal transport in the gas phase. Thus, the total thermal conductivity λ is represented by the sum of the individual contributions:

$$\lambda = \lambda_{\text{solid}} + \lambda_{\text{rad}} + \lambda_{\text{gas}}, \qquad (3.1)$$

with λ_{solid} the thermal conductivity of the solid backbone, λ_{rad} the radiative conductivity, and λ_{gas} the thermal conductivity of the pore gas.

While the transport via the solid backbone can be minimised by high porosity and a high fraction of dead ends (i.e., mass of the backbone that does not contribute to the heat transport), the thermal conductivity along the gas phase can efficiently be suppressed by reducing the pore size to values below the mean free path of the gas molecules, l_{gas}, at a given gas pressure p_{gas} and temperature T (Kistler, 1935). For air at ambient temperature T_0 and pressure ($p_{gas,0}$), the mean free path $l_{gas,0}$ is about 70 nm. For other conditions the mean free path can be calculated by the following relationship:

$$l_{gas}\left(p_{gas}, T\right) = l_{gas,0} \sqrt{\frac{T}{T_0} \frac{p_{gas,0}}{p_{gas}}}. \tag{3.2}$$

With respect to radiative thermal transport, one has to distinguish between IR-optically thin and thick materials. In optically thin media, characterised by

$$\tau(T) = e^*(T) \cdot \rho \cdot d \ll 1, \tag{3.3}$$

the radiative transport depends on the radiative properties of the macroscopic boundaries of the insulation and thus is not a material but rather a system property. Here, τ is the optical thickness of the material; $e^*(T)$, its temperature-dependent specific extinction coefficient; ρ, its density; and d, the layer thickness of the material under investigation.

In optically thick materials ($\tau \gg 1$), heat transport is characterised by the radiative thermal conductivity:

$$\lambda_{rad}(T) = \frac{16}{3} \cdot \frac{n^2 \cdot T_r^3}{\rho \cdot e^*(T)}, \tag{3.4}$$

with n the effective index of refraction and T_r the mean radiative temperature of the material. This implies that the contribution due to this transport path can be reduced by providing high specific extinction coefficients. The latter depend on both the chemical composition as well as the size and shape of the radiation-reducing component integrated in the material under investigation.

Depending on the chemical composition of the aerogel, specific extinction coefficients between some 10 m²/kg and 1,500 m²/kg (carbon aerogels) are observed. If the intrinsic radiative extinction of the bare aerogel is not sufficient, organic modifications (that can be pyrolysed) and additives (pigments) can be introduced (see e.g., Lu *et al.*, 1992; Lee *et al.*, 1995; Schwertfeger and Schubert, 1995).

In some cases aerogels are available as granules rather than monoliths (see e.g., Figure 3.2.). The thermal transport of a bed of granules is given by a superposition of the heat transfer along the gas phase in the intergranular pores and gas, solid, and radiative transport within the granules (Reim *et al.*, 2004). The effective solid-phase contribution strongly depends on the external load impounded onto the bed of granules, thus changing the contact area at the granule/granule interface; coupling between the gaseous thermal conductivity in between the granules and the heat transport along the granule contact points is usually another factor to be taken into account.

Low-temperature insulations: Since low-temperature thermal insulations are usually applied under vacuum, the pore size is not a crucial factor for the heat transport. For temperatures less than 10 K, thermal conductivities below $4 \cdot 10^{-5}$ W/m·K can be provided under vacuum with silica aerogels having densities of about 0.07 kg/m^3 (Scheuerpflug, 1992). At 173 K the same sample reaches a thermal conductivity of about $2 \cdot 10^{-3}$ W/m·K. Organic aerogels of similar densities are expected to show even lower thermal conductivities as the intrinsic conductivity of the corresponding solid phase is smaller in amorphous organic materials (Lu *et al.*, 1992).

The fact that aerogels are not only excellent thermal insulations but also self-supporting materials that can be provided as transparent solids opens new opportunities for insulations needed in special fields of low-temperature research.

In low-temperature applications that do not operate under high vacuum, aerogels have additional advantages compared with open porous organic foams or fibre insulations since the gaseous contribution to the overall heat transport is significantly suppressed for pores smaller than the mean free path in the gas phase. This is why silica aerogel (with an outer gold layer as a radiation shield) was used as lightweight thermal insulation in the NASA Mars rover (Jones, 2006).

Ambient temperature insulations: At ambient temperatures silica aerogel–based materials (granules, monoliths, and composites) are interesting insulation material since they are non-flammable, translucent, or even transparent materials with thermal

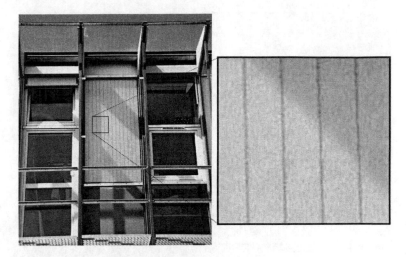

Figure 3.4 Facade element consisting of polycarbonate double-bridge panels filled with silica aerogel granules (© ZAE Bayern). See also Colour Insert.

conductivities down to $13 \cdot 10^{-3}$ W/m·K. Therefore they allow combining thermal insulation properties with transparency or translucency, for example, for use in daylighting systems. Examples are silica aerogel granules incorporated in polycarbonate double-bridge panels (Figure 3.4). The low thermal conductivities are a result of the high porosity in combination with pore sizes smaller than 100 nm.

In a bed of granules, the thermal properties of the individual granules are superimposed by thermal transport along the gas phase in the intergranular pores; due to typical spacings of a few 100 microns to millimeter, the gaseous transport in those voids is governed by the properties of the free gas under the given conditions. Figure 3.5 shows the thermal conductivity for a bed of silica granules as a function of gas pressure. The two steps indicate the intergranular voids (step at low gas pressure) and the pores within the aerogel granules (step near ambient pressure). The height of the low-pressure step is proportional to the volume fraction of intergranular voids in the system and can therefore be controlled by the shape and size distribution of the granules.

Although granular aerogels do not show the low thermal conductivities that are found for monoliths, they provide advantages

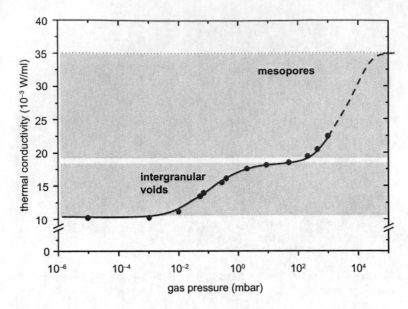

Figure 3.5 Thermal conductivity of a bed of silica granules vs. gas pressure (Reim *et al.*, 2004). See also Colour Insert.

upon processing and thus production costs on one side and are more flexible in terms of serving complex insulation shapings. One example here is the use of opacified silica aerogel powder as thermal insulation material in an evacuated double-wall container, which was applied by BMW as thermal insulation for a latent heat storage system for the motor to improve its efficiency.

At given porosity and pore size, the lowest thermal conductivities are achieved for organic aerogels. The reason here is the low intrinsic thermal conductivity of the organic backbone. For densities of about 160 kg/m^3 values as low as 0.012 W/m·K are feasible (Lu *et al.*, 1992). Organic aerogels are synthesised from phenolic reactants (resorcinol, melamine, phenol) in combination with aldehydes, polyurethane, or cellulose (see e.g., Biesmans *et al.*, 1998; Rigacci *et al.*, 2004; Hoepfner *et al.*, 2008).

Organic-based aerogels can be applied up to about 200 to 250°C, while inorganic materials are thermally stable up to at least 450°C. Depending on the structure and the composition, metal oxide–based (e.g., mullite-type) aerogels with long-term thermal stability can be provided for temperatures up to about 1,000°C.

High-temperature thermal insulations: Thermal insulations for temperatures well above 1,000°C (e.g., process chambers for graphitisation) are often applied under inert atmospheres or even under vacuum. Under these conditions carbon aerogels that are derived by pyrolysis of organic aerogels are excellent alternatives to conventional carbon fibre–based thermal insulations. The main advantages of carbon aerogels lie in their small pore sizes (thus suppressing the gaseous thermal transport) and in particular in their huge specific radiative extinction.

Although consisting – just like the commercial fibre felts – of pure carbon, the spacial dispersion of the carbon in the insulation is far more effective in aerogels. Thus with respect to radiative transport, they behave like monoliths composed of highly dispersed carbon black. The latter is well known as an extremely effective pigment for the blocking of infrared radiation. Figure 3.6 shows the direct

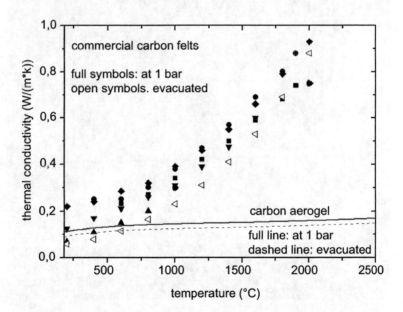

Figure 3.6 Thermal conductivity of commercial carbon felts and an ambient pressure-dried carbon aerogel (Wiener, 2009; Wiener *et al.*, 2009).

comparison of commercial carbon felt–based thermal insulations for temperatures up to about 2,500°C and a carbon aerogel. While below 1,000°C the thermal conductivities for both materials are in the same range, at higher temperatures the aerogel is showing a significantly higher insulation performance.

Interesting systems can also be provided by combining carbon and SiC phases in the insulation. The SiC phases can be incorporated by silica introduced during or after the sol-gel process. At temperatures above 1,200°C and non-oxidising atmosphere, the silica phase is reacting with neighboring carbon to form SiC; simultaneously, part of the carbon is burned off due to a reaction with locally provided oxygen-containing products from the reaction of silica and carbon to SiC (Figure 3.7). As a result composites with low thermal conductivity at given porosity can be provided. For example, values of 0.1 to 0.15 W/m·K, even at temperatures well above 1,000°C, can be achieved for a porosity of 85% (Reichenauer *et al.*, 2008).

HV	det	mag	WD	pressure	HFW	——————— 2 µm ———————
5.00 kV	LFD	50 000 x	6.9 mm	60 Pa	5.97 µm	Quanta FEG

Figure 3.7 SEM micrograph showing a carbon/SiC composite (© ZAE Bayern).

Figure 3.8 Laminated carbon aerogel tile; left: photo; right: SEM image of fleece/tile interface (© ZAE Bayern). See also Colour Insert.

Carbon aerogels can be produced as monolithic tiles that can be further customised by machining or milling; for better handling the tiles can be laminated with a carbon fleece or fabric, thus providing a component with high bending stiffness (Figure 3.8).

To reduce abrasion, thin layers of pyrolytic carbon or SiC can be applied on the outer surface in addition.

Customised shapes can be produced by moulding of the organic precursors. Figure 3.9 shows that even complex shapes are easily being achieved.

Recently, it was shown that aerogels can be synthesised with a controlled gradient in their structure (Figure 3.10) and/or density (Hemberger *et al.*, 2009), thus providing continuous gradients instead of layered insulation systems that are less affected by thermal stress.

Figure 3.9 Carbon aerogel with the shape of a Lego™ brick. Shaping was performed by moulding of the organic precursor (© ZAE Bayern). See also Colour Insert.

Figure 3.10 Structurally graded carbon aerogel. Here a 20 mm wide gradient (indicated by the arrow) in terms of pore size was provided, showing a continuous transition in pore size from 60 nm to 6 microns (see also Hemberger *et al.*, 2009).

3.1.2.2 Components for energy storage

Electrically conductive and chemically inert carbon aerogels with specific surface areas up to about 2,000 m²/g and tailorable pore structures are excellent candidates for storage systems such as electrochemical double-layer capacitors (EDLC) (Mayer *et al.*, 1993; Pekala *et al.*, 1998; Conway 1999; Probstle *et al.*, 2000; Hwang and Hyun, 2004).

In those devices electrical charge is electrostatically stored at the carbon/pore interface. The principle is the same as in conventional capacitors, where the capacitance C is given by:

$$C = \frac{\varepsilon \cdot \varepsilon_0 \cdot A}{d},$$ (3.5)

with ε and ε_0 the dielectric constants, A the area of the capacitor plates, and d the distance between the two electrodes.

The main difference between EDLCs and conventional parallel-plate capacitors is the huge specific interface available and the small distance between the layers of opposite charge: d in EDLCs is defined by the so-called Helmholtz layer and is on the order of nanometer rather than microns to millimeters.

EDLCs are components for all applications that require fast charging or discharging rates, and thus access to the total storage capacity has to be feasible on a time scale of seconds; examples are uninterruptible power supply (UPS), regenerative breaking of automotives, the autarkic control of the slip control in wind energy systems, and the buffering of fast fluctuations in photovoltaic (PV) systems.

Usually EDLCs are made using activated charcoal powder in combination with a highly conductive additive and a binder to provide thin layer electrodes for high specific power. The same concept can be used with powderised activated carbon aerogels prepared by grinding carbon aerogel granules or by sol-gel synthesis of small spherical carbon particles (Figure 3.11 and Scherdel *et al.*, 2009). Alternatively, a thin layer of electrically well-conductive carbon aerogel composite can be provided by carbon fibre–reinforced, activated carbon aerogels. These materials are derived by infiltration of an organic or carbon fibre fleece or fabric with the solution for the organic carbon aerogel precursor (Wang *et al.*, 2001). After controlled reaction to a well-defined organic gel backbone integrated in the fibre matrix, the composite is dried at ambient conditions and eventually pyrolysed. The resulting composite can easily be cut to the final electrode shape.

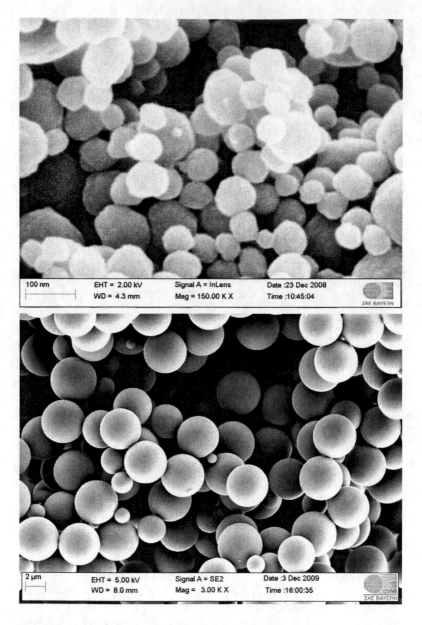

| 100 nm | EHT = 2.00 kV | Signal A = InLens | Date :23 Dec 2008 | |
| | WD = 4.3 mm | Mag = 150.00 K X | Time :10:45:04 | ZAE BAYERN |

| 2 μm | EHT = 5.00 kV | Signal A = SE2 | Date :3 Dec 2009 | |
| | WD = 8.0 mm | Mag = 3.00 K X | Time :16:00:35 | ZAE BAYERN |

Figure 3.11 Carbon aerogel particles of different sizes (© ZAE Bayern).

In contrast to typical activated carbon, aerogels provide a carbon backbone with high chemical purity, thus reducing degradation effects, and tailorable meso- and macroporosity.

The latter can be exploited to specifically optimise the transport properties in the electrolyte phase and thus the power density of the device. Further advantages are the high electrical conductivities of aerogel-based carbons (prepared without the use of an organic binder), in particular when attached by chemical bonding to well-conducting additives (e.g., graphene, Worsley *et al.*, 2010) or conventional carbon fibres that simultaneously allow for the synthesis of mechanically stable thin layers.

One of the current trends is to microscopically combine EDLC-type and battery-type functionalities to achieve high energy densities at power densities well above those of pure battery systems. Hereby a phase containing redox functions is introduced into the system by, for example, infiltration of the carbon matrix with a manganese oxide precursor; the latter is then converted in situ to Mn_3O_4 nanofibres attached to the carbon aerogel skeleton (Lin *et al.*, 2011) or by a carbon-assisted reaction into a MnO_2–coating on the carbon backbone (Fischer *et al.*, 2007), (see Figure 3.12).

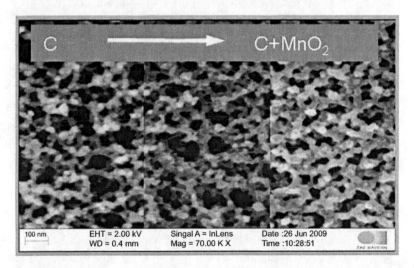

Figure 3.12 Carbon aerogel MnO_2 hybrid electrode for electrical energy storage (© ZAE Bayern).

Alternatively, co-precipitation can be applied to provide such composites (Li *et al.*, 2006).

Recently carbon aerogels have been identified as promising matrices for hosting complex metal hydrides applied for hydrogen storage. The role of the matrix is hereby to accommodate the metal hydride in such a way that the storage kinetics is strongly enhanced as the matrix restricts the particle size of the hydride and provides a well-accessible and highly dispersed metal hydride storage component (see e.g., Gross *et al.*, 2008).

3.1.2.3 Catalysts supports

Catalysts are most efficient when they are provided well dispersed and accessible to the phase in which the respective reaction is to be catalysed. In some cases the catalyst support plays an additional role in terms of the overall efficiency of the catalyst: for example, wetting properties might be crucial whenever liquid phases are present or formed upon the reaction. Furthermore a favourable mass transport or increased electrical conductivity of the matrix may enhance the efficiency of the reaction at the catalyst.

In this context aerogels as support for catalysts provide many new options compared with conventional supports such as charcoal, carbon black, or ceramics (Colmenares *et al.*, 1991; Hirashima *et al.*, 1997; Hirashima *et al.*, 1998; Ferino *et al.*, 2000; Fabrizioli *et al.*, 2002; Alie *et al.*, 2004; Dunn *et al.*, 2005; Jayne *et al.*, 2005; Wang and Ro, 2005; Hao *et al.*, 2008; Bali *et al.*, 2009; Hao *et al.*, 2009).

In addition to tailored mesoporosity, graded porous structures or macropores for high overall throughput can easily be implemented in aerogels. For the integration of the metal catalysts on the inner surface of carbon aerogels, three different strategies can be applied (Moreno-Castilla and Maldonado-Hodar, 2005):

- the introduction of a metal precursor (e.g., metal salts) in the initial solution for the synthesis of the organic carbon aerogel precursor (e.g., Job *et al.*, 2007),
- the use of functionalised phenolic groups in the sol-gel process, which later enhance the ion exchange in the organic (aero)gel, and

- the deposition of metal precursors in the organic or carbon aerogel via the liquid or the vapour phase.

Although several of the papers on catalysts supported on carbon aerogels still report on using supercritical drying for the processing of the organic aerogel precursor, drying at ambient conditions can likely be applied in most cases unless the supercritical drying also has an impact on the catalyst dispersion.

Similar approaches as given above for carbon aerogel supports can also be used for inorganic aerogel-based catalyst hosts.

Investigations of the catalytic activity in selected combustion, chemical synthesis processes, and growth of carbon nanotubes (CNTs) (Piao and Li, 2003) as well as in electrodes for fuel cells (Baker *et al.*, 2004; Moreno-Castilla and Maldonado-Hodar, 2005) reveal a high potential of aerogels as catalyst supports.

3.1.2.4 Other energy-related fields of application

Due to their well-designable open-porous structure, aerogels with very high or specifically tailored low porosity may also be introduced in other fields of energy-related applications such as

- gas separation and filtering,
- membranes for osmosis,
- thermodiffusion,
- hosts for fusion targets, or
- components in thermoelectric devices.

In all cases their versatility in terms of providing functionalised composites, thin layers or monolithic moulds in combination with mesopores architecture, and well-accessible surface area make aerogels superior to other types of porous materials.

3.1.3 *Problems to be Solved for a Broad Introduction of Aerogels in Energy-Related Applications*

Although aerogels have a huge potential as next-generation materials in energy technologies, currently only very few large companies are serving energy-related applications with aerogel-based products. Examples are Cabot Inc. providing silica aerogel granules and boards as well as coated textile membranes, Aspen Aerogels Inc. producing flexible blankets based on fibre/aerogel composites as thermal insulations for different temperature ranges, Rockwool entering the market with Aerowolle®, a fibre/silica aerogel composite board for thermal insulation, and PowerStor® advertising EDLC devices with carbon aerogels electrodes.

Reasons herefore are the fact that until a few years ago, nano-materials were considered model systems in fundamental research rather than future commercial products. When the awareness of nanomaterials in industries began, a lot of the focus was drawn to particular materials such as nanotubes and nanoparticles rather than to monolithic highly porous components such as aerogels.

The few companies producing aerogels are still working on de-veloping more economic processing techniques since, especially the drying step represents (in particular for non-granular components) still a challenge in terms of costs.

In addition, current low-density inorganic aerogel-based mate-rials are often problematic in terms of dust release upon handling and customising or in the application itself; this is another point that producers are currently addressing.

Although organic and carbon aerogels usually do not have similar disadvantages in terms of handling, they also have barriers to overcome to be commercialised on a broader base. The US patents filed in the late 1980s and mid-1990s (Pekala, 1988; Kaschmitter, 1992; Droege, 1997) made carbon aerogels less attractive to new industrial players. In addition, it was not until a few years ago that the first carbon aerogel tiles with a cross-sectional area of 10 to 10 cm^2 were prepared crackfree by ambient pressure drying on the lab scale (Wiener, 2009).

Recently interesting alternative phenol-based routes in com-bination with convective drying have been developed (Scherdel and Reichenauer, 2009), that make carbon aerogels promising candidates for future high-temperature thermal insulation systems. So far, the furnace industry, a more conservative business, is still hesitant to apply new types of materials even if they show superior performance at similar costs compared with carbon felts or foams.

Progress has also been made very recently in terms of new processes for the production of particles that either are only microporous or possess a porous structure in the meso- to macropore range. These materials could be used in adsorption-related or electrochemical applications as well as systems that require highly efficient pigments.

3.1.4 *Conclusions*

Sol-gel-derived porous solids and composites provide almost unlimited options for functional materials specifically designed to meet the demands of different energy-related applications. While in the past the term *aerogel* was generally related to costly, hard-to-handle materials, the new generation of aerogels is increasingly changing this picture, making its way from high-end space applications to a down-to-earth tool box of engineers to be used as components in devices and systems. Although quite a few active producers exist in the United States and Asia, aerogels still lack a broader pool of producers, in particular in Europe that could provide the base for more technical and economic competitiveness.

Bibliography

Alie, C., Ferauche, F., Heinrichs, B., Pirard, R., Winterton, N., and Pirard, J. P. (2004). Preparation and characterization of xerogel catalyst microspheres, *J. Non-Cryst. Solids*, 350, 290–298.

Baumann, T. F. (2010). synthesis of graphene aerogel with high electrical conductivity, *J. Am. Chem. Soc.*, 132(40), 14067–14069.

Babic, B., Kaluderovic, B., Vracar, L., and Krstajic, N. (2004). Characterization of carbon cryogel synthesized by sol-gel polycondensation and freeze-drying, *Carbon*, 42(12–13), 2617–2624.

Baker, W. S., Long, J. W., Stroud, R. M., and Rolison, D. R. (2004). Sulfur-functionalized carbon aerogels: a new approach for loading high-surface-area electrode nanoarchitectures with precious metal catalysts, *J. Non-Cryst. Solids*, 350, 80–87.

Bali, S., Huggins, F. E., Huffman, G. P., Ernst, R. D., Pugmire, R. J., and Eyring, E. M. (2009). Iron aerogel and xerogel catalysts for Fischer-Tropsch synthesis of diesel fuel, *Energy Fuels*, 23(1), 14–18.

Baxter, J., Bian, Z., Chen, G., Danielson, D., Dresselhaus, M. S., Fedorov, A. G., Fisher, T. S., Jones, C. W., Maginn, E., Kortshagen, U., Manthiram, A.,

Nozik, A., Rolison, D. R., Sands, T., Shi, L., Sholl, D., and Wu, Y. (2009). Nanoscale design to enable the revolution in renewable energy, Energy Environ. Sci., 2(6), 559–588.

Biesmans, G., Randall, D., Francais, E., and Perrut, M. (1998). Polyurethane-based aerogels for use as environmentally acceptable super insulants, *Cell. Polym.*, 17(1), 17–30.

Bisson, A., Rigacci, A., Lecomte, D., Rodier, E., and Achard, P. (2003). Drying of silica gels to obtain aerogels: phenomenology and basic techniques, *Drying Technol.*, 21(4), 593–628.

Brandhuber, D., Torma, V., Raab, C., Peterlik, H., Kulak, A., and Hüsing, N. (2005). Glycol-modified silanes in the synthesis of mesoscopically organized silica monoliths with hierarchical porosity, *Chem. Mater.*, 17(16), 4262–4271.

Brinker, C. J. and Scherer, G. W. (1990). Sol-Gel Science: The Physics and Chemistry of Sol-Gel Processing, Academic Press.

Brinker, C. J., Sehgal, R., Hietala, S. L., Deshpande, R., Smith, D. M., Loy, D. and Ashley, C. S. (1994). Sol-gel strategies for controlled porosity inorganic materials, *J. Membr. Sci.*, 94, 85–102.

Colmenares, C., Connor, M., Evans, C., and Gaver, R. (1991). Photoactivated heterogeneous catalysis on aerogels, *Euro. J. Solid State Inorg. Chem.*, 28, 429–432.

Conway, B. E. (1999). Electrochemical Supercapacitors - Scientific Fundamentals and Technological Applications, Kluwer Academic Publishers/Plenum Press, New York.

Davis, P. J., Brinker, C. J., and Smith, D. M. (1992). Pore structure evolution in silica-gel during aging drying. 1. Temporal and thermal aging, *J. Non-Cryst. Solids*, 142(3), 189–196.

Droege, M. W. (1997). WO1999001502A1.

Dunn, B. C., Cole, P., Covington, D., Webster, M. C., Pugmire, R. J., Ernst, R. D., Eyring, E. M., Shah, N., and Huffman, G. P. (2005). Silica aerogel supported catalysts for Fischer-Tropsch synthesis, *Appl. Catal., A*, 278(2), 233–238.

Ebert, H.-P. (2011). Thermal properties of aerogels, in: Aerogels Handbook, Eds: Aegerter, M. A., Leventis, N., and Koebel, M., Springer (Berlin), 537–562.

Fabrizioli, P., Burgi, T., and Baiker, A. (2002). Environmental catalysis on iron oxide-silica aerogels: selective oxidation of NH_3 and reduction of NO by NH_3, *J. Catal.*, 206(1), 143–154.

Ferino, I., Casula, M. F., Corrias, A., Cutrufello, M. G., Monaci, R., and Paschina, G. (2000). 4-methylpentan-2-ol dehydration over zirconia catalysts prepared by sol-gel, *Phys. Chem. Chem. Phys.*, 2(8), 1847–1854.

Fischer, A. E., Pettigrew, K. A., Rolison, D. R., Stroud, R. M., and Long, J. W. (2007). Incorporation of homogeneous, nanoscale MnO_2 within ultraporous carbon structures via self-limiting electroless deposition: implications for electrochemical capacitors, *Nano Lett.*, 7(2), 281–286.

Fricke, J. (1994). Thermal Transport in Nanostructured Porous Materials and Their Optimization as Thermal Superinsulators, 10th Intl. Heat Transfer Conference, Brighton, England, August 14–18, 1994.

Gross, A. F., Vajo, J. J., Van Atta, S. L., and Olson, G. L. (2008). Enhanced hydrogen storage kinetics of $LiBH_4$ in nanoporous carbon scaffolds, *J. Phys. Chem. C*, 112(14), 5651–5657.

Haereid, S., Anderson, J., Einarsrud, M. A., Hua, D. W., and Smith, D. M. (1995). Thermal and temporal aging of tmos-based aerogel precursors in water, *J. Non-Cryst. Solids*, 185(3), 221–226.

Hao, Z. G., Zhu, Q. S., Jiang, Z., Hou, B. L., and Li, H. Z. (2009). Characterization of aerogel Ni/Al_2O_3 catalysts and investigation on their stability for CH_4-CO_2 reforming in a fluidized bed, *Fuel Proc. Technol.*, 90(1), 113–121.

Hao, Z. G., Zhu, Q. S., Jiang, Z., and Li, H. Z. (2008). Fluidization characteristics of aerogel Co/Al_2O_3 catalyst in a magnetic fluidized bed and its application to CH_4-CO_2 reforming, *Powder Technol.*, 183(1), 46–52.

Hemberger, F., Weis, S., Reichenauer, G., and Ebert, H.-P. (2009). Thermal transport properties of functionally graded carbon aerogels, *Int. J. Thermophys.*, 30(4), 1357–1371.

Hirashima, H., Kojima, C., and Imai, H. (1997). Application of alumina aerogels as catalysts, *J. Sol-Gel Sci. Technol.*, 8(1–3), 843–846.

Hirashima, H., Kojima, C., Kohama, K., Imai, H., Balek, V., Hamada, H., and Inaba, M. (1998). Oxide aerogel catalysts, *J. Non-Cryst. Solids*, 225(1), 153–156.

Hoepfner, S., Ratke, L., and Milow, B. (2008). Synthesis and characterisation of nanofibrillar cellulose aerogels, *Cellulose*, 15(1), 121–129.

Hüsing, N. and Schubert, U. (2005). Aerogels, Wiley-VCH Verlag GmbH & Co. KGaA, Weinheim.

Hwang, S. W. and Hyun, S. H. (2004). Capacitance control of carbon aerogel electrodes, *J. Non-Cryst. Solids*, 347(1–3), 238–245.

Jayne, D., Zhang, Y., Haji, S., and Erkey, C. (2005). Dynamics of removal of organosulfur compounds from diesel by adsorption on carbon aerogels for fuel cell applications, *Int. J. Hydrogen Energy*, 30(11), 1287–1293.

Job, N., Pirard, R., Vertruyen, B., Colomer, J.-F., Marien, J., and Pirard, J.-P. (2007). Synthesis of transition-metal doped carbon xerogels by cogelation, *J. Non-Cryst. Solids*, 353(24–25), 2333–2345.

Jones, S. M. (2006). Aerogel: space exploration applications, *J. Sol-Gel Sci. Technol.*, 40(2–3), 351–357.

Kaschmitter, J. L. M., Steven, T., and Pekala, R. W. (1992). US05789338A, 1992.

Kistler, S. S. (1932). Coherent expanded aerogels, *J. Phys. Chem. A*, 36(1), 52–64.

Kistler, S. S. (1935). The relation between heat conductivity and structure in silica aerogel, *J. Phys. Chem.*, 39(1), 79–86.

Kocklenberg, R., Mathieu, B., Blacher, S., Pirard, R., Pirard, J. P., Sobry, R., and Van den Bossche, G. (1998). Texture control of freeze-dried resorcinol-formaldehyde gels, *J. Non-Cryst. Solids*, 225(1), 8–13.

Lee, D., Stevens, P. C., Zeng, S. Q., and Hunt, A. J. (1995). Thermal characterization of carbon-opacified silica aerogels, *J. Non-Cryst. Solids*, 186, 285–290.

Li, J., Wang, X. Y., Huang, Q. H., Gamboa, S., and Sebastian, P. J. (2006). A new type of MnO_2 center dot $xH(2)O$/CRF composite electrode for supercapacitors, *J. Power Sources*, 160(2), 1501–1505.

Lin, Y.-H. Wei, T.-Y., Chien, H.-C., and Lu, S.-Y. (2011). Manganese oxide/carbon aerogel composite: an outstanding supercapacitor electrode material, *Adv. Energy Mater.*, 1(5), 901–907.

Lu, X., Arduini-Schuster, M. C., Kuhn, J., Nilsson, O., Fricke, J., and Pekala, R. W. (1992). Thermal conductivity of monolithic organic aerogels, *Science*, 255(5047), 971–972.

Lu, X., Wang, P., Arduinischuster, M. C., Kuhn, J., Buttner, D., Nilsson, O., Heinemann, U., and Fricke, J. (1992). Thermal transport in organic and opacified silica monolithic aerogels, *J. Non-Cryst. Solids*, 145(1–3), 207–210.

Mayer, S. T., Pekala, R. W., and Kaschmitter, J. L. (1993). The aerocapacitor - an electrochemical double-layer energy-storage device, *J. Electrochem. Soc.*, 140(2), 446–451.

Moreno-Castilla, C. and Maldonado-Hodar, F. J. (2005). Carbon aerogels for catalysis applications: an overview, *Carbon*, 43(3), 455–465.

Pekala, R. W. (1988). US04997804A.

Pekala, R. W., Farmer, J. C., Alviso, C. T., Tran, T. D., Mayer, S. T., Miller, J. M., and Dunn, B. (1998). Carbon aerogels for electrochemical applications, *J. Non-Cryst. Solids*, 225(1), 74–80.

Piao, L. Y. and Li, Y. D. (2003). Structure of carbon nanotubes from decomposition of methane on an aerogel catalyst, *Acta Physico-Chimica Sinica*, 19(4), 347–351.

Poco, J. F., Coronado, P. R., Pekela, R. W., and Hrubesh, L. W. (1996). A rapid supercritical extraction process for the production of silica aerogels, 7.

Prakash, S. S., Brinker, C. J., Hurd, A. J., and Rao, S. M. (1995). Silica aerogel films prepared at ambient-pressure by using surface derivatization to induce reversible drying shrinkage, *Nature*, 374(6521), 439–443.

Probstle, H., Saliger, R., and Fricke, J. (2000). Electrochemical investigation of carbon aerogels and their activated derivatives, *Charact. Porous Solids V*, 128, 371–379.

Reichenauer, G. (2004). Thermal aging of silica gels in water, *J. Non-Cryst. Solids*, 350, 189–195.

Reichenauer, G. (2008). Aerogels. Kirk-Othmer Encyclopedia of Chemical Technology, John Wiley & Sons.

Reichenauer, G., Wiener, M., and Ebert, H.-P. (2008). DE102008037710A1.

Reim, M., Reichenauer, G., Korner, W., Manara, J., Arduini-Schuster, M., Korder, S., Beck, A., and Fricke, J. (2004). Silica-aerogel granulate - structural, optical and thermal properties, *J. Non-Cryst. Solids*, 350, 358–363.

Rigacci, A., Einarsrud, M. A., Nilsen, E., Pirard, R., Ehrburger-Dolle, F., and Chevalier, B. (2004). Improvement of the silica aerogel strengthening process for scaling-up monolithic tile production, *J. Non-Cryst. Solids*, 350, 196–201.

Rigacci, A., Marechal, J. C., Repoux, M., Moreno, M., and Achard, P. (2004). Preparation of polyurethane-based aerogels and xerogels for thermal superinsulation, *J. Non-Cryst. Solids*, 350, 372–378.

Scherdel, C., Scherb, T., and Reichenauer, G. (2009). Spherical porous carbon particles derived from suspensions and sediments of resorcinol-formaldehyde particles, *Carbon*, 47, 2244–2252.

Scherdel, C. and Reichenauer, G. (2009). Carbon xerogels synthesized via phenol-formaldehyde gels, *Microporous Mesoporous Mater.*, 126, 133–142.

Scheuerpflug, P. (1992). Tieftemperatureigenschaften von SiO_2-Aerogelen (PhD), Bayerische Julius-Maximilians-Universität, Würzburg.

Schwertfeger, F. (1991). Darstellung hydrophober Aerogele über den Sol-Gel-Prozeß (Diplom), Julius-Maximilians-Universität, Würzburg.

Schwertfeger, F. and Schubert, U. (1995). Generation of carbonaceous structures in silica aerogel, *Chem. Mater.*, 7(10), 1909–1914.

Tamon, H., Ishizaka, H., Yamamoto, T., and Suzuki, T. (1999). Preparation of mesoporous carbon by freeze drying, *Carbon*, 37(12), 2049–2055.

Wang, J., Glora, M., Petricevic, R., Saliger, R., Proebstle, H., and Fricke J. (2001). Carbon cloth reinforced carbon aerogel films derived from resorcinol formaldehyde, *J. Porous Mater.*, 8(2), 159–165.

Wang, C. T. and Ro, S. H. (2005). Nanocluster iron oxide-silica aerogel catalysts for methanol partial oxidation, *Appl. Catal. a-Gen.*, 285(1–2), 196–204.

Wiener, M. (2009). Synthese und Charakterisierung Sol-Gel-basierter Kohlenstoff-Materialien für die Hochtemperatur-Wärmedämmung (PhD), Bayerische Julius-Maximilians-Universität, Würzburg.

Wiener, M., Reichenauer, G., Braxmeier, S., Hemberger, F., and Ebert, H.-P. (2009). Carbon aerogel-based high-temperature thermal insulation, *Int. J. Thermophys.*, 30(4), 1372–1385.

Wiener, M., Reichenauer, G., and Fricke, J. (2003). Structural Changes upon Gelation and Aging of Carbon Aerogel Precursors, HASYLAB am Deutschen Elektronen-Synchrotron DESY.

Worsley, M. A., Pauzauskie, P. J., Olson, T.Y., Biener J., Satcher, J. H., Jr., and Baumann, T. F. (2010). synthesis of graphene aerogel with high electrical conductivity, *J. Am. Chem. Soc.*, 132(40), 14067–14069.

3.2 Energy Sources and Conversion

3.2.1 *Dye Solar Cells*

Dr. Claus Lang-Koetz, Dr. Andreas Hinsch, and Dr. Severin Beucker
Germany

This article reports on the current state of technology and market development of DSC technology, the scaling-up of glass-based DSC modules, and the development of a production technology. It focuses on the activities of the research project **ColorSol – Sustainable Product Innovation through Dye Solar Cells**, which was funded by the German Federal Ministry for Research from 2006 to 2008.

The objective of the project was to develop the DSC technology to the stage of application, outline the requirements for a pilot production, and realise demonstration applications. The project consortium consisted of the medium-sized companies BGT Bischoff Glastechnik, Pröll, Engcotec, and IoLiTec and the research institutes Fraunhofer Institute for Solar Energy Systems ISE, Fraunhofer Institute for Industrial Engineering IAO, and Borderstep Institute for Innovation and Sustainability (also see www.colorsol.de).

3.2.1.1 DSC technology and its application

DSCs belong to a new group of PV devices. A DSC is a photoelectrochemical system. The DSC technology was originally invented by Michael Grätzel and Brian O'Regan at the École Polytechnique Fédérale de Lausanne (O'Regan and Grätzel, 1991) and is also known as the 'Grätzel cell'.

In a DSC, an organic dye is used to convert light into electrical energy. This dye is embedded in nanocrystalline titanium dioxide electrodes. The design of a typical DSC is depicted in Figure 3.13: the dye, titanium dioxide, and the electrolyte are located in a hollow space between two glass plates coated with different functional layers.

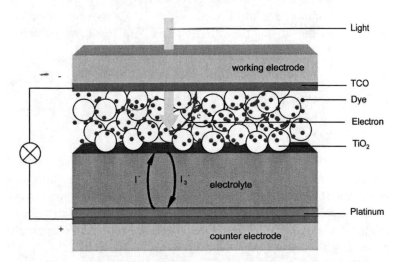

Figure 3.13 Principal design of a DSC (Fraunhofer ISE). See also Colour Insert.

DSCs have so far reached solar efficiencies up to 11.2% under standard testing conditions; 8.2% have been reported for small modules by Sharp Solar.

Fraunhofer ISE has developed semi-transparent DSC modules with internal interconnection (meander type) in the size of 30 × 30 cm using glass frit-sealing technology. In a second development step, the scaling-up of the module to the size of 60 × 100 cm was undertaken.

As of now, the efficiency on the total module area for the 30 × 30 cm module developed by Fraunhofer ISE is 3.5%, with 4.2% on the active module area. A module efficiency of approximately 4–5% on the total area can be realised in the short term with this concept. Much higher efficiencies are foreseen for the future (Chiba *et al.*, 2006).

DSCs have the following advantages:

(1) To manufacture them, comparably low-priced materials and relatively simple production technology can be used.
(2) A wide area of design options is possible: The colour of the cells can be varied from dark red to amber, and the transparency can be changed by printing light-scattering layers into the cells (see Figure 3.14).
(3) Compared with silicon-based PV, DSCs are more sensitive to diffuse light irradiation and less sensitive to higher ambient temperature (see Hinsch *et al.*, 2009).

These advantages make them an ideal technology for building integrated photovoltaics (BIPV) (Hinsch *et al.*, 2007).

3.2.1.2 Characteristics of DSC modules

At low light illumination, DSCs show a storage effect. On an outdoor testing facility of Fraunhofer ISE, manufactured dye solar modules were measured during different radiation intensities and over several months. As expected, below 300 W/m^2, that is, below 0.3 sun intensities, a higher performance was shown than for comparable solar modules out of crystalline silicon per W_p (Watt peak).

Furthermore, the dependency of the relative module efficiency on the orientation of the solar modules (angle of incidence) was analysed (see Figure 3.15 below). It was shown that the dependency of the relative performance is much lower for dye solar modules: this is due to the good optical coupling onto the photoactive layer. At an angle of incidence of 50 degrees or higher, a relative advantage of 10% could be measured compared with a standard solar module.

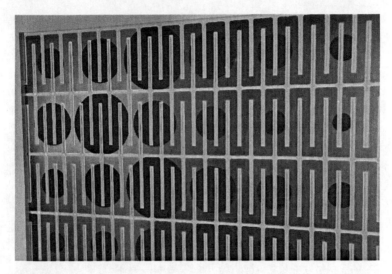

Figure 3.14   DSC prototype manufactured in the project ColorSol. See also Colour Insert.

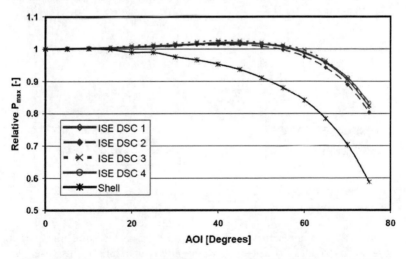

Figure 3.15   Comparison AOI; DSC vs. Si module (Lang-Koetz *et al.*, 2009).

These results show that DSCs are especially well suited for the application in building facades which are often not optimally aligned to the sun and where naturally a lower solar irradiation can be found with a higher angle of incidence.

3.2.1.3 Manufacturing steps for DSC modules

Glass frit-sealed dye solar modules (30 × 30 cm) can be manufactured by screen printing (see Figure 3.16 and a detailed description in Sastrawan *et al.*, 2006).

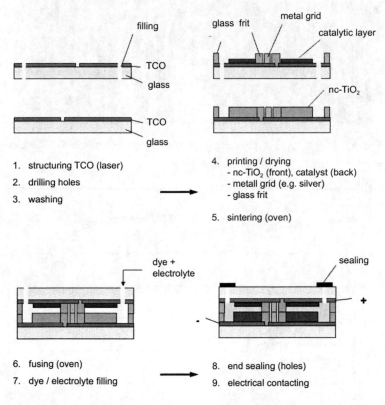

Figure 3.16 Manufacturing scheme of glass frit-sealed, semi-transparent dye solar modules as developed at the Fraunhofer ISE. In this case, two TCO(SnO$_2$:F)-coated glass substrates are used. See also Colour Insert.

Only nine principal production steps are needed, requiring no vacuum technology or cleanroom facilities. Most of the necessary manufacturing equipment is available from the glass manufacturing and processing industry. The processing itself uses glass substrates, making it ideal for manufacturing in façade glass production.

The glass frit sealing used in this approach provides high-quality sealing. An important advantage of the manufacturing process is that dye and purified electrolytes are introduced after the sealing of the modules through filling holes.

Each module is internally sixfold serially interconnected via screen-printed 'Z' contacts. Module prototypes can be laminated onto architectural glass.

In a further step experiences gained from the manufacturing of 30×30 cm modules were used to realise the scaling-up of modules to 60×100 cm (see Figure 3.17).

3.2.1.4 Industrial production for DSC modules

In the project ColorSol it was shown that DSCs can be manufactured in an industrial environment. Industrial production consists of the following steps:

First, glass with a conductive layer (TCO) is pre-treated and structured, and then the functional layers (TiO_2, ZrO_2, platinum, silver, glass frit) are printed on the glass by multiple steps of screen printing. Special printing pastes are required: Such pastes were further developed in the project ColorSol by Pröll KG with Fraunhofer ISE. Printing processes and paste production were scaled-up and printing parameters optimised.

The subsequent manufacturing steps were successfully scaled up by the company BGT Bischoff Glastechnik on the basis of knowledge from Fraunhofer ISE: screens with different layouts were manufactured and different screen fabrics tested in order to reach the required precision with respect to layer thickness and sharpness of edges.

The printed players are dried, and then a sintering process step follows at 580°C, during which the organic compounds of the pastes vanish. This is followed by a fusing process step in an oven at even higher temperature, at which the melting point of the glass

Figure 3.17 Filling of a DSC module (60 × 100 cm). See also Colour Insert.

frit is reached. That leads to the sealing of the glass plates and the formation of a set of individual glass chambers with a defined distance between the glass plates. The empty glass modules are then filled with dye and electrolytes, and the openings in the glass plates are sealed with a UV hardening glue. An additional lamination step can be required, if safety glass is needed, for example, when applying the DSC modules in an overhead area.

Most described production steps can be realised by adapting existing processes and technologies from the glass industry. However, filling and sealing of the 60 × 100 cm DSC modules can only be conducted in a reasonable time span with low manual effort and with a high reproducibility when they are automated.

For this reason, a filling machine was developed by Fraunhofer IAO and Fraunhofer ISE in order to perform the following process steps: 1) leak-testing individual cells with compressed air, 2) colouring with a dye solution, 3) cleaning, 4) filling with an electrolyte, 5) cleaning the tube system, and 6) sealing the cells. The machine was built as an industrial prototype based mainly on standard components. It can perform steps 1 through 5 automatically and has an appliance to support step 6. Steps for testing and process controls are included in the machine design (see Figure 3.18).

In the final step of production, the DSC modules are contacted and built into a frame system.

In the project ColorSol, the company Pröll KG has successfully scaled up the production of the most important printing pastes. Different dispersing methods were evaluated and printing features optimised.

As an electrolyte, an ionic liquid – a material consisting fully of ions existing a liquid aggregate state below 100°C – is used. The company IoLiTec has developed appropriate ionic liquids, adapted them, and successfully produced them in a micro-reactor.

Figure 3.18 Pictures of the developed industrial prototype of a DSC-filling machine. See also Colour Insert.

3.2.1.5 Application scenarios for future DSC products

BIPV account so far for a rather small portion of solar applications. For a number of reasons, this is likely to change in the coming years.

First, a number of international surveys verified the potential of BIPV (IEA, 2002; EuPD, 2008; NanoMarkets, 2008). The overall potential in the 14 most important worldwide markets sums up to 23,270 km^2, including roofs and facades with suitable orientation to the sun (IEA, 2002).

Second, a number of European countries provide additional funding for BIPV applications. France and Italy, for example, offer higher feed-in tariffs for applications in facades or roof integation.

Third, new technologies like semi-transparent DSCs offer a new variety of construction materials. Until now only glass-based DSCs allow a high degree in freedom for the design and the optical appearance of building integrated modules.

Aside from the new design properties offered by DSCs, they also dispose of features like better heat resistance or higher yields in indirect and scattered light conditions, which makes them specifically applicable in BIPV applications (see above).

An important objective of the projects ColorSol was to evaluate customer needs and future markets for BIPV applications of DSCs. For that purpose, a cooperation with leading architects from the field of BIPV was set up. As a first step, existing market surveys on thin-film technologies were analysed and expert interviews were accomplished to identify attributes (e.g., degree of solar efficiency, general requirements from building codes) that are important for future applications. In a second step, a series of workshops was conducted to evaluate different applications of BIPV and specific requirements for new PV products such as DSC modules. The results were summed up in a third step in three scenarios that were further detailed and completed with precise technical, legal, and economical development goals for future products and applications.

The three specified scenarios for DSC modules in BIPV are:

- Building facades in different forms, for example, as facing with and without air space,
- PV glazing, for example, in glass roofs, structural glazing and similar constructions, and
- Windows as part of PV-rooftop installations.

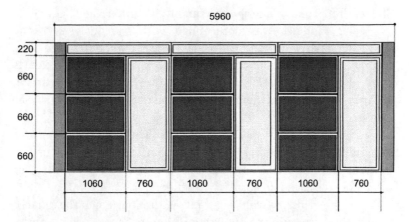

Figure 3.19 Design of a facade for a mobile DSC architectural demonstrator (red areas: dye solar modules). See also Colour Insert.

The most important outcome of the workshops series was that many applications of BIPV can be financially competitive in new buildings if the total system cost like substituted façade material substitute energy and feed in tariffs for solar energy are taken into account.

To show future applications of DSC modules, a façade demonstrator is being developed in a new project (see Figure 3.19) (funding provided by the Ministry for the Environment of the State of Baden-Württemberg, contract BUT011).

In addition to the requirements, cost calculations for future products were performed and a road map for a pilot production of the 60 × 100 cm modules was developed. The calculated costs were compared with data from the production of thin-film PV that are competing technologies. The calculation of the estimated future production and product costs show that the DSC modules can be produced at costs comparable with other thin-film technologies.

3.2.1.6 Environmental impact

In order to regard principles for an environmentally friendly technology development, the following activities were performed in the project ColorSol:

The entire primary energy consumption within the lifecycle of dye solar modules was determined with the analysis of the cumulated energy demand (CED). For retrieving the required data, the experience of the ColorSol project partners was used, experts were questioned, existing literature was reviewed, a comparison with similar processes was conducted, and data from LCA databases was retrieved.

As the most important indicator, the energy payback time was determined. This number shows the time in which the use of dye solar modules has supplied as much energy as was used for production, use, and disposal of a module.

Different scenarios were regarded: within the standard scenario, the parameters represent the state of the technology and realistic installation conditions (module efficiency of 5%, installation in Germany, integration into a façade directed towards the south). A lifetime of 10 years was assumed.

For these parameters, an energetic payback time of 29.8 months was calculated. It can be reduced to 17 months by an installation of the modules in Southern Europe, more efficient production processes, and increased performance; however, less favourable conditions could lead to a rise up to 40 months (Rist and Pastewski, 2008).

This energy payback time can be seen as competitive when compared with other solar cells (see Table 3.1).

The glass component has the highest resource consumption and recycling relevance, also since it is the component with the highest mass in the DSC system. The chosen electrolyte is not toxic

Table 3.1 Comparison of energy paypack time for different solar cell technologies (values for DSC: own analysis, other values: from Quaschning, 2009)

Type of solar cell	Energy payback time (in months)
mono-Si	55
poly-Si	38
amorphous	28
CIS	15
DSCs	17–40

or mutagenic; this was shown in tests conducted by a certified laboratory for the ColorSol project. Others have reported that no health or safety risks were identified for the ruthenium dye so far (results of the Ames test – a standard test to assess the mutagenic potential of chemical compounds – in de Vries *et al.*, 2000). It could be shown that to decrease the energy payback time, not only alignment of the modules to the sun and the solar efficiency of the modules, but also transport distance within the value chain has the highest impact. Hence, they should be closely regarded when further optimising the technology (Rist and Pastewski, 2008).

Recycling of dye solar modules is possible by using mechanical-chemical processes. A possible recycling procedure was determined: The modules have to be disassembled and components broken up. The resulting pieces can be treated with sodium hydroxide to remove the electrolyte. Titanium dioxide and dye can be separated with an ethanole solution, filtration, sedimentation, and distillation. The remainder consists of pieces of glass with very thin prints of silver, platinum, and zirconium dioxide layers. These can be treated further or simply be used in a normal glass-recycling process. The selection of respective processes depends on an economical-ecological assessment for each specific case. For a final statement, further practical tests are required.

3.2.1.7 Conclusions and outlook

DSC modules with glass frit sealing are particularly promising for the application in facades and BIPV. Market evaluation based on surveys and expert interviews indicates potentials for their application in new building materials for BIPV.

Having started in the publicly funded research project ColorSol, a network of research institutes and small and medium-sized companies has successfully been created to scale up modules and to realise steps of a pilot production in an industrial environment. This network will be further developed within follow-up activities with the objective to integrate additional partners that can bring in necessary competency for the setup of a pilot production line and commercialisation.

It could be shown that performance and efficiency of DSCs under outdoor conditions are comparable to silicon modules with specific advantages for DSC modules in indirect and scattered light conditions (Beucker *et al.*, 2009). It was also shown that DSC modules react robustly under partial shading, since the cell acts like an internal bypass diode in a serial interconnection (Hinsch *et al.*, 2009). This also makes DSCs suitable for BIPV applications.

For the further characterisation and certification of DSCs and modules, the efficiency of the cell has to be determined under modified conditions. Due to the nature of DSCs as an electrochemical solar technology, the response time of the cells and modules is longer than that of other PV technologies. Thus, a flash lamp test as used in the efficiency calibration of standard PV modules cannot be applied. Also, a standard norm for accelerated testing has still to be developed by adapting existing IEC norms for PV modules.

In the future, a focus will be put on the development of pilot products and a pilot market for façade applications of dye solar modules. For the commercialisation of dye solar modules, a number of activities such as reliability and performance tests under standard test conditions (STC) and pilot projects for façade integration have to be performed.

Acknowledgments

The work described in this paper has been funded by the German Federal Ministry of Education and Research (BMBF) (project ColorSol: contract 01 RI 05211).

Bibliography

Beucker, S., Hinsch, A., Brandt, H., Veurman, W., Flarup J., K., Lang-Koetz, C., and Stabe, M. (2009). Scaling-up of glass based dsc-modules for applications in building integrated photovoltaics, Proc. 34th IEEE Photovoltaic Spec. Conf., Philadelphia, Pennsylvania, June 7–12, 2009.

Chiba, Y., Islam, A., Watanabe, Y., Komiya, R., Kodie, N., and Han, L. (2006). Dye-sensitised solar cells with conversion efficiency of 11.1%, *Jap. J. Appl. Phys.*, 45, 638–640.

de Vries, J. G., Scholtens, B. J. R., Maes, I., Grätzel, M., Winkel, S., Burnside, S., Wolf, M., Hinsch, A., Kroon, J. M., Ahlse, M., Tjerneld, F., Ferrero, G., Bruno, E., Hagfeldt, A., Bradbury, C., Carlsson, P., Pettersson, H., Verspeek-Rio, C. M., and Enninga, I. C. (2000). Negative Ames-test of cis-di(thiocyanato)-N, N'-bis(4,4'-dicarboxy-2,2'-bipyridine)Ru(II), the sensitizer dye of the nanocrystalline TiO_2 solar cell, *Sol. Energy Mater. Sol. Cells*, 60(1), 43–49, Elsevier Science B.V.

EuPD. (2008). The German Photovoltaic Market 2007/08-From Sales to Strategic Marketing, Report, Bonn, Germany. Available at http://www.eupd-research.com/.

Hinsch, A., Putyra, P., Würfel, U., Brandt, H., Skupien, K., Drewitz, A., Einsele, F., Gerhard, D., Gores, H., Hemming, S., Himmler, S., Khelashvili, G., Nazmutdinova, G., Opara Krasovec, U., Rau, U., Sensfuss, S., Walter, J., and Wasserscheid, P. (2007) Developments towards low-cost dye solar modules, in: European Commission, Joint Research Centre: 22nd European Photovoltaic Solar Energy Cenference and Exhibition 2007. Proceedings, *Milano, Italy*.

Hinsch, A., Brandt, H., Veurman, W., Hemming, S., Nittel, M., Würfel, U., Putyra, P., Lang-Koetz, C., Stabe, M., Beucker, S., and Fichter, K. (2009). Dye solar modules for façade applications: recent results from project ColorSol, Solar Energy Mater. *Solar Cells*, 93, 820–824.

International Energy Agency. (2002). Potential for Building Integrated Photovoltaics, Report IEA PVPS Task 7, NET Ltd., St. Ursen, Switzerland.

Lang-Koetz, C., Hinsch, A., and Beucker, S. (2009). Farbige Solarzellen. Grundlage für eine attraktive gebäudeintegrierte Photovoltaik, Erneuerbare Energien, March 2009, *Hannover Germany*, 51–55.

NanoMarkets. (2008). Thin Film Photovoltaics Markets: 2008 and Beyond, Report Item Number: Nano-054, Glen Allen, VA, USA. Available at http://www.nanomarkets.net/.

O'Regan, B. and Grätzel, M. (1991). A low-cost, high-efficiency solar cell based on dye-sensitized colloidal TiO_2 films. *Nature*, 353(6346), 737–740. doi:10.1038/353737a0.

Quaschning, V. (2009). Overview from different literature sources, compiled by Prof. Dr. Volker Quaschning, homepage with information on renewable energies and climate protection, Berlin. Available at http://www.volker-quaschning.de/ (last access: October 2009).

Rist, M. and Pastewski, N. (2008). Die Technologie der Farbstoffzelle, Zeitschrift Sonne, Wind & Wärme, 32(14), 130–133.

Sastrawan, R., Beier, J., Belledin, U., Hemming, S., Hinsch, A., Kern, R., Vetter, C., Petrat, F.M., Prodi-Schwab, A., Lechner, P., and Hoffmann, W. (2006). New interdigital design for large area dye solar modules using a lead-free glass Frit sealing, *Prog. Photovolt.: Res. Appl.*, 14(8), 697–709.

Toyoda, T., Sano, T., Nakajima, J., Doi, S., Fukumoto, S., Ito, A., Tohyama, T., Yoshida, M., Kanagawa, T., Motohiro, T., Shiga, T., Higuchi, K., Tanaka H., Takeda, Y., Fukano, T., Katoh, N., Takeichi, A., Takechi, K., and Shiozawa, M. (2004). Outdoor performance of large scale DSC modules, *J. Photochem. Photobiol.* A, 164(1–3), 203–207.

3.2.2 Nanoscale Thermoelectrics – a Concept for Higher Energy Efficiency?

Dr. Harald Böttner, Jan König
Fraunhofer Institute for Physical Measurement Techniques, Germany

Cars, machines, and power stations are not fuel efficient; on average they utilise about one-third of the energy supplied to them. The rest, in the form of heat, is dissipated. If it were possible to recover the large amounts of waste heat energy, the energy balance of combustion processes would be significantly more favourable. The quest for methods leads to a technology that has been known for well over a hundred years – thermoelectrics. Thermoelectric energy harvesting, the collection of waste heat from combustion processes, is currently enjoying high popularity and has triggered a worldwide research race for higher efficiencies, inspired by visions such as the car as a mobile mini-power station.

If it were possible, at least partly, to power vehicle electronics with exhaust heat, fuel savings of some percentage, as claimed by the automotive industry, could be achieved. The heart of thermoelectric systems is commonly massive and bulk-like. Currently the object of intensive research is thermoelectric materials with sophisticated atomic, so-called nanocsale internal structures.

Here the simplest physical basis for thermoelectrics will be summarised, and the rationale for the physical concept behind nanoscale thermoelectrics will shortly be mentioned. Present different nanoscale concepts will be presented, and challenges for technical, mature thermoelectric systems based on nanoscale materials will be highlighted.

3.2.2.1 Introduction

Thermoelectrics has been defined as the science and technology of thermoelectric generation (production of electricity) and refrigeration (production of cold).

It is energy conversion due to the Seebeck coefficient (thermal energy converted to electrical energy: battery) and due to the Peltier coefficient (electrical current converted in heat transport: refrigeration). Several semiconductors are well known as suitable thermoelectric materials. The value of a material in thermoelectric conversion is measured by its so-called figure of merit (Z), which is more conveniently expressed in its dimensionless form ZT, where T is absolute temperature. Z is $S^2 \times \sigma/\lambda$, with S = Seebeck coefficient [μV/K], σ = electrical conductivity [Ω cm], and λ = thermal conductivity [W/m·K]. The devices are thus known as thermogenerators and Peltier coolers. Figure 3.20 shows these two modes of operation for thermoelectric devices.

Traditional applications for thermogenerators are well known, including, for example, spacecraft applications, pacemakers, or

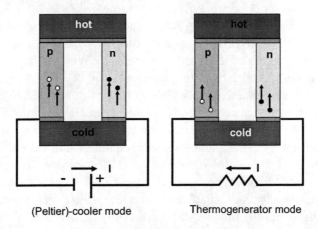

(Peltier)-cooler mode   Thermogenerator mode

Figure 3.20 Thermoelectric devices: Mode of operation. See also Colour Insert.

stand-alone electrical power stations. Established products in the area of Peltier coolers are various cooling applications even in consumer products like mobile wine cabinets. As far as thermo-generators are concerned, thermoelectric technologies possess as a unique feature environmentally friendly solutions to regain part of the inevitable loss of heat in various productions of energy or during industrial fabrication processes.

3.2.2.2 Initial concepts of nanoscale thermoelectrics

The question whether a thermal converter is economically viable depends, on the one hand, on its thermoelectric power, meaning what the voltage is at a certain temperature difference. On the other hand, efficient thermoelectric converters should conduct electricity well, while being poor heat conductors.

The temperature gradient is as large as possible when the material is heated on one side. Here, the natural laws of physics put a spoke in the material scientist's wheel: from basic physics it is impossible to obtain high electrical conductivity simultaneously with low thermal conductivity.

Electrical conductivity and thermal conductivity are connected in that σ is tied to λ through the Wiedemann-Franz relationship $\lambda = L\sigma T$, where the Lorentz number L is the proportional factor between λ and σ (L is equal to $L_0 = 2.45 \times 10^{-8}$ V^2/K^2 for metals and strongly degenerated semiconductors; for semiconductors L depends on the band structure). Typical values of thermal conductivity for a good TE material are $\lambda < 2$ Wm^{-1}K^{-1}.

Starting in 1992 two complementary concepts had been proposed to overcome the classical situation of interconnected electrical and thermal conductivity. Hicks and Dresselhaus (Dresselhaus *et al.*, 1992) proposed, similar to the concept of the quantum well design for semiconductor lasers, quantum well systems (zero-dimensional [0D], one-dimensional [1D], and two-dimensional [2D]) for thermoelectric materials to enhance, due to the quantum design, the Seebeck coefficient and the electrical conductivity, keeping, hopefully, the thermal conductivity untouched.

This will enhance the power factor that is the numerator ($S^2\sigma$) of the thermoelectric figure of merit ZT.

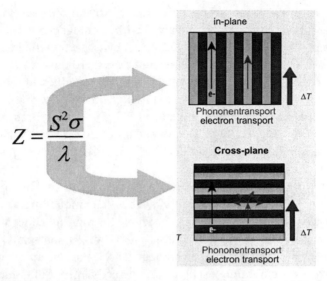

Figure 3.21 First nanoscale thermoelectric concepts from 1992. See also Colour Insert.

To the same time Venkatasubramanian (Dresselshaus *et al.*, 1992) at the Research Triangle Institute (RTI) proposed to use nano-engineered multilayer structures known as superlattices (SLs). The scatter of mid- to long-wavelength phonons at each layer boundary of such SLs reduces the thermal conductivity, keeping, also hopefully, the electrical conductivity untouched. In 2001 (Venkatasubramanian *et al.*, 1992) it was published that through Bi_2Te_3/Sb_2Te_3 SLs, ZT values for p-type material of $\sim$2.4 were possible.

It should be mentioned that up to now, this data is unique, that is, no other lab in the world was able to reproduce these breakthroughs for the field of thermoelectrics. Figure 3.21 represents in a simple sketch both complementary concepts.

3.2.2.3 Current concepts of nanoscale thermoelectrics

A nanostructure has one of its crucial dimensions in the order of 1 – 100 nm. Nanostructures have various forms such as quantum dots, nanowires, SLs, and nanoparticles incorporated in bulk materials, for example, nanocomposites.

Theoretical approaches predicted a strong ZT-rise in low-dimensional structures compared with homogeneous material systems. This could be achieved by a power factor enhancement due to carrier quantum confinement (Dresselshaus *et al.*, 1992) or by a reduction of the lattice thermal conductivity due to phonon confinement and/or enhanced phonon-scattering mechanisms (Venkatasubramanian *et al.*, 1992; Hicks and Dresselhaus, 1993). Even higher ZT-values exceeding the calculated values for 2D transport are predicted for 1D structures such as quantum wires.

The formation of such nanoscale materials can be fabricated by different techniques. In this overview these techniques are arranged in bottom-up and top-down concepts. Bottom-up concepts are understood as nanoscale-structured (0D, 1D, or 2D) materials grown on substrates by thin-film techniques. These bottom-up nanoscale structures are not only useful for the principal proof of theoretical concepts, they have also practical relevance as there are two new companies (Nextreme 2005 USA and Micropelt 2006 Germany) founded in the last decade which produce thin-film devices for special cooling applications and small-energy self-sufficient sensor systems.

The field of energy harvesting in automotive application, for example, is addressed by the top-down concept, including nanostructured bulk materials like multiphase materials or nanocomposites.

3.2.2.3.1 Bottom-up concepts   Top-down concepts cover the classical thin-film and nanowire approach. These concepts can be classified by the electronic density of states in a 2D quantum well, 1D nanowire, and 0D quantum dot approaches.

Quantum dots: The formation of quantum dots is possible using, for example, physical-vapour-deposition techniques like molecular beam epitaxy (MBE). If strain is introduced in heteroepitaxy by lattice mismatch, it can lead to an energetically more favourable island growth and the formation of self-organised quantum dots. This behaviour was observed in lead chalcogenide systems by several groups (Springholz *et al.*, 1998; Nurnus *et al.*, 2003; Harman *et al.*, 2000).

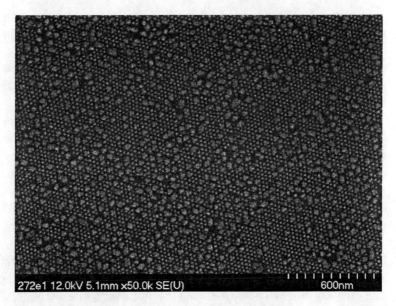

Figure 3.22 SEM image of self-organised PbSe/PbTe quantum dots (Nurnus *et al.*, 2003).

A typical scanning electron microscopy (SEM) image of self-organised PbSe/lead telluride (PbTe) quantum dots is shown in Figure 3.22.

Harman and his group grew with the same method Tl-doped p-type quantum dot SLs consisting of $PbSe_{0.98}Te_{0.02}$ and PbTe layers. Se-rich quantum dots are formed at the interfaces of the layers.

They discovered in the year 2000 a $ZT = 0.9$ at 300 K, which is more than a factor of two higher than the ZT of PbTe used for thermoelectric applications (Harman *et al.*, 2000). With Bi-doped (n-type) quaternary PbSnSeTe/PbTe QDSL samples, they improved the ZT value to 2 at 300 K two years later (Harman *et al.*, 2002). Since PbTe has its optimum thermoelectric performance at higher temperatures, an n-type PbSeTe/PbTe quantum dot SL reached a maximum ZT of 3 at 550 K (Harman *et al.*, 2005). This ZT rise is said to be caused both by an enhancement of the thermopower compared with bulk PbTe and by a reduction of the lattice thermal conductivity λ_{latt}. The lattice thermal conductivity is reduced to one-fourth of the lattice thermal conductivity of bulk-like PbTe.

A similar reduction by as much as a factor of 3 compared with the corresponding bulk alloy is also measured in the InGaAs:ErAs system (Kim *et al.*, 2006). Nanoparticles of ErAs formed in an InGaAs matrix with a dimension of some nanometres. The self-ordered particles along the <110> directions really reduce the thermal conductivity.

The concept of quantum dots is obviously useful for reducing the lattice thermal conductivity and improving thermoelectrically the figure of merit. Nevertheless there is to our knowledge no company up to now trying to fabricate commercial thermoelectric thin-film modules based on quantum dot SLs.

This bottom-up concept is transferred into bulk materials, resulting also in an improvement of the figure of merit.

Nanowires: The main idea for nanowires is to use a 1D system to separate phonons from electronic transport. This could be experimentally achieved by reducing the diameter of the nanowires. If the diameter of a nanowire is in the same length scale like the phonon-free path of a material or is even smaller, then the lattice thermal conductivity should be dramatically reduced.

This approach is more interesting for materials with a large lattice thermal conductivity, which means a large phonon mean free path as it is for silicon. The mean free path for phonons is about 300 nm compared with a mean free path of about 100 nm for electrons in highly doped samples. Different groups work on the fabrication of Si nanowires to reduce the thermal conductivity for the improvement of the thermoelectric properties of silicon. It was observed for Si nanowires with a diameter of 22 nm that the thermal conductivity is smaller than 1 W/m·K, which is more than a 100-fold reduction compared with the bulk value (Li *et al.*, 2003). Si nanowires with a rough surface and diameters of about 50 nm exhibit also thermal conductivity two orders of magnitude lower (Hochbaum *et al.*, 2008). Regarding thermoelectric properties, it is reported that these rough Si nanowires have Seebeck coefficients and electrical resistivities that are similar to bulk Si. This yields a *ZT* value of 0.6 at room temperature.

Also in other systems a reduction in the thermal conductivity is observed. For example, the thermal conductivity in single-crystalline

PbTe nanowires could be reduced at 300 K by almost a factor of 2 for a nanowire diameter of 182 nm compared with the thermal conductivity of bulk PbTe (Roh *et al.*, 2010). This is a huge reduction for such a large diameter, considering the low lattice thermal conductivity of bulk PbTe single crystals ($\sim$2 W/m·K).

These individual PbTe nanowires are grown by a chemical vapour transport method in contrast to an aqueous electro-less etching method for the growth of the rough Si nanowires in wafer-scale arrays. The advantage of vertical-aligned Si nanowires in a wafer scale fabrication method is also the more practical.

Another approach to fabricating practicable nanowires is the electrochemical deposition of thermoelectric materials in anodic alumina (AAO) templates. These AAO templates contain self-organised arrays of aligned pores. These pores are vertically oriented with a diameter between 10 to some hundreds of nanometres. These holes can be filled with different thermoelectric materials, that is, Bi or Bi_2Te_3. An empty and a Bi_2Te_3-filled AAO template is shown in Figure 3.23. The main problem of this approach is the too high thermal conductivity of the template. Further development is focused on the replacement of the template.

SLs: SL-based thermoelectric materials (see section 3.2.2.2 and also Figure 3.21) mostly focus on perfect, that is, epitaxial, multi-layer, systems and equipment accordingly. Although the fabrication of these multilayer materials is commonly cost intensive and, in

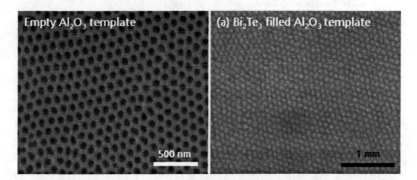

Figure 3.23 An empty and a Bi_2Te_3-filled AAO template. (K. Nielsch, private communications).

particular, extremely time consuming – growth rates are typically in the order of ~ 1 μm/h – one of these systems, based on V–VI compounds, achieved technical ripeness. Under V–VI compounds the Bi_2Te_3-based materials have the highest ZT around room temperature. Consequently and finally, successful efforts were started to develop suitable epitaxial systems for this material system. Venkatasubramanian and co-workers (2001) reported about Bi_2Te_3-based SLs grown by metallorganic chemical vapour deposition (MOCVD) on GaAs substrates. High-quality multilayers for p-type material (Bi_2Te_3/Sb_2Te_3 multilayers) and n-type material, that is, $Bi_2Te_3/Bi_2(Se,Te)_3$, were grown. Abrupt SL interfaces with minimal inter-layer mixing were achieved.

The reported ZT data of 2.4 for p-type Bi_2Te_3/Sb_2Te_3 and of 1.7 for n-type $Bi_2Te_3/Bi_2(Se,Te)_3$ SLs are clearly the most important results in the context of SL multilayer thermoelectrics. So far, the largest ZT enhancement was reached in the cross-plane (c-direction) direction. For this anisotropic material the cross-plane direction displays inferior thermoelectric properties if compared with the in-plane direction (a-direction).

It could be demonstrated that the cross-plane lattice thermal conductivity dropped down significantly to $\lambda_L \sim 0.22$ W/m·K, while reducing the periodicity for the p-type Bi_2Te_3/Sb_2Te_3 SLs to ~ 6 nm. Moreover, a significant decrease for the thermal conductivity was measured in plane for n-type, $Bi_2Te_3/Bi_2(Se_{0.12}Te_{0.88})_3$, SLs with a 10 nm layer thickness period (Nurnus *et al.*, 2000). The thermal conductivity was reduced to 65% compared with the values for homogenous Bi_2Te_3. Until now, there is virtually no data available on the stability of such thin SLs against cation (p-material) and anion (n-material) interdiffusion. For possible applications the thermal and long-term stability of the SLs is of extraordinary importance.

On the basis of the RTI results, a complete technology was established to process epitaxial multilayers for miniaturised thermoelectric devices in the well-known standard vertical design. In contrast to standard devices, the leg height is reduced here down to only some micrometres. This enables these devices to have the unique feature of massive enhanced cooling power densities up to ~ 100 W/cm^2, which is about two decades higher compared with standard thermoelectric cooling devices, the Peltier-coolers.

During 2005 this technology led to the foundation of the US start-up company Nextreme. Although the initial business case for those miniaturised thermoelectric devices was Peltier cooling, nowadays these devices are used in a rapidly growing market for energy harvesting in autonomous sensor systems with included wireless data transfer. The typical converted electrical energy here is in the range of mW to μW.

Other epitaxial thermoelectric multilayer systems like IV–VI SLs and mixed IV–VI/V–VI SLs are exclusively of academic interest without any expectation of future practical use. Details of these SLs are discussed in Nurnus *et al.*, (2006).

3.2.2.3.2 Top-down concepts
Top-down concepts cover bulk materials with built-in nanostructures. The existing nanostructured bulk materials and concepts behind can be classified as bulk quantum dots, nanocomposites, and multiphase SL bulk materials.

Bulk quantum dots: The bottom-up concept of quantum dots is transferred into bulk materials. The so-called 'LAST', or Pb-Sb-Ag-Te, and 'SALT', or Na-Sb-Pb-Te, materials are such bulk quantum dot materials. These materials are based on PbTe compounds with a small amount of $AgSbTe_2$ and $NaSbTe_2$, respectively, or compositional variations. It was shown by transmission electron microscopic analysis that compositional modulations on the 1–10 nm scale, including endotaxially dispersed quantum dots, exist (Hsu *et al.*, 2004). These inclusions result in a reduction of the thermal conductivity almost similar as in the bottom-up concept. A *ZT* of $\sim$2.2 ($\sim$1.7) at about 800 K (700 K) for the n-type $AgPb_mSbTe_{2+m}$ (m = 10 and 18) was reported.

The synthesis from the melt is very difficult because of the strong temperature-dependent cooling process and the spinodale decomposition as well as the interaction of Ag and Sb with the PbTe matrix.

Ag and Sb act in PbTe as amphoteric dopants, which means that they can cause p- and n-type conduction, depending on the Pb/Te ratio. Different approaches fail to reproduce the good thermoelectric properties of this material. The reason is said to be the preparation conditions. That's why different synthesis can help in a large fabrication method of this material. This approach uses a rapid cooling technique and a kind of hot pressing. These techniques are discussed in detail next.

3.2.2.4 Nanocomposite bulk materials

As a short outlook, a ZT of 1.5 is reached at 675 K for the same composition fabricated by this approach (Zhou *et al.*, 2008). A compatible p-type material is the so-called SALT system consisting of PbTe with a small amount of $NaSbTe_2$. It is interesting that almost the same ratio of matrix material (PbTe) to precipitates ($NaSbTe_2$) of about 20:1 in the p-type $NaPb_mSbTe_{2+m}$ (m = 20) leads also to a huge reduction of the thermal conductivity and to a high figure of merit of about 1.6 at 675 K (Hoang *et al.*, 2005). The overall p-type conduction can be explained by the doping character of Na. Na behaves as an efficient acceptor dopant in PbTe.

An extended miscibility gap and thus a spinodale decomposition exist also in the PbTe-PbS system. Coherent nanoinclusions of PbS are determined in PbTe and $Pb_{0.95}Sn_{0.05}Te$. In the system $(Pb_{0.95}Sn_{0.05}Te)_{0.92}(PbS)_{0.08}$, a reduction in the lattice thermal conductivity of about 30% to that of bulk PbTe at room temperature was observed. These quantum dot–like nanoinclusions result in a ZT of 1.5 at 650 K (Sootsman *et al.*, 2006).

The bulk quantum dot approach could be a possibility for large-scale synthesis of high-performance thermoelectric materials.

Nanocomposite bulk materials: Nanocomposite bulk materials can be classified into two groups. The first group contains materials consisting of a matrix of an already good thermoelectric material in which nanostructures are inserted.

Different inert nanoparticles such as fullerene C60 (Shi *et al.*, 2004), ZrO_2 (He *et al.*, 2007), and TiO_2 (Xiong *et al.*, 2009) were added to the matrix of, for example, skutterudites via mechanical mixing and resulted in a reduction of the lattice thermal conductivities. The inclusion of fullerene and ZrO_2 in skutterudites showed only a small success. An increase of 16% for the figure of merit ZT ~1.1 at 850 K is obtained for a sample of $Ba_{0.22}Co_4Sb_{12}$ with 0.4 vol.% TiO_2 (Xiong *et al.*, 2009).

The incorporation of C_{60} in Bi_2Te_3-based composites has been found useful for lowering the thermal conductivity but was without any success in increasing the figure of merit (Gothard *et al.*, 2010). On the other hand a *ZT* value of 1.25 was reported at 420 K for hydrothermally synthesised nanotubes embedded in a Bi_2Te_3 matrix. This composite showed n-type conduction (Xiao *et al.*, 2006).

The second group consists of nanostructured and compacted composites. Typically the green body is reduced to small pieces with (nanoscale) dimensions compacted by pressing and sintering.

A new processing route to obtain nanostructured materials regarding an improvement of *ZT* is the usage of rapid cooling techniques and compaction of the obtained nanostructured powders (Tang *et al.*, 2007; Ebling *et al.*, 2007).

Nanoscaled inclusions can be obtained due to the very high cooling rates because there is not enough time for the formation of larger crystals. The so-called 'melt spinning' is such a rapid cooling technique. High cooling rates of about 10^{-3}–10^{-7} K/s could be reached by dropping liquid melt on a fast-rotating, nitrogen-cooled copper wheel. This technique is known to be used to fabricate amorphous metal. So it is ideal to create inclusions in nanoscale dimensions due to the very short time scale of the cooling process.

The challenge is to keep these nanostructures during the compaction process. Spark plasma sintering (SPS) is seen as one of the most promising techniques for thermoelectric nanocomposites due to the very fast heating rates.

The SPS process is a kind of a modified hot-pressing technique. A high, pulsed electrical current is applied on the pressed powders, which heats up the powder in a very short time. It is expected that during sintering, a local electric arc is emerged between the particles and temporally limited high temperature and pressure occur. Another advantage is that thin oxide layers around the nanoparticles can be broken and a good electrical contact could be achieved.

This processing route results in p-type Bi_2Te_3 ingots with a *ZT* value of 1.35 at 300 K (Tang *et al.*, 2007). Ribbons with a layered nanostructure were fabricated by the melt-spinning technique. The SPS-compacted samples also show a layered strucuture with a layer size of 10–40 nm in width. It was shown that the nanostructure obtained by the melt-spinning process can be preserved after the SPS process and that the nanostructure influences the electric and thermal transport properties. A minimum lattice thermal conductivity of 0.58 W/m·K was reached at 300 K.

The highest ZT value for a bulk p-type Bi_2Te_3-based material was reported by Poudel *et al.*, (2008). The material was prepared by ball milling followed by direct-current hot pressing. This material is a single-phase material made of a composition of nanoparticles and macroparticles. A ZT of 1.4 at 100°C was reported. The ZT value enhancement for this system is caused by a reduction of the phonon thermal conductivity, while maintaining a comparable power factor to that of the bulk Bi_2Te_3-based material.

The fabrication of nanostructured composite is also interesting for the bulk quantum dot materials which are difficult to fabricate. $AgPb_mSbTe_{2+m}$ with m $=$ 18 (the same composition as mentioned in section 3.2.2.3.1) showed a ZT value of about 1.5 at 673 K. This sample was produced using mechanical alloying and SPS (Zhou *et al.*, 2008).

The nanostructured composite concept was applied to boron-doped, p-type $Si_{95}Ge_5$, and an enhanced ZT of about 1 at 800°C was reported, which is similar to that of large-grained $Si_{80}Ge_{20}$ alloys (Zhu *et al.*, 2009). The most important improvement is the reduction of the Ge content and therefore the reduction in material cost.

The materials were prepared by mechanical alloying using a ball mill technique. The nanopowders were then compacted by using a direct-current hot-pressing method for rapid compaction (almost similar to the above-mentioned SPS process).

The ZT improvement originates from the high density of nanograin interfaces which act as phonon-scattering centers, reducing the thermal conductivity. The same nanostructuring strategy is also applied to P-doped n-type $Si_{80}Ge_{20}$ using the same fabrication process (Wang *et al.*, 2008). For this material a ZT of about 1.3 at 900°C is reported.

The nanostructured composite concept seams to work very well and could be a possibility for large-scale thermoelectric material production. One issue that arises in such a nanoscale material is the long-term stability of nanograins against the growth of grains at elevated temperatures.

Multiphase SL bulk materials: Another concept which originates from the bottom-up concepts is the multiphase SL bulk materials.

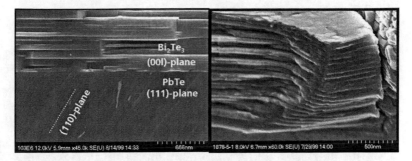

Figure 3.24 (a) Epitaxial grown Bi_2Te_3 on top of a 111-oriented PbTe surface and (b) epitaxially grown IV–VI/V_2–VI_3 SLs (Nurnus *et al.*, 2000).

It was presented in 2000 that two different semiconductors with good thermoelectric properties, as it is for Bi_2Te_3 and PbTe, can be grown epitaxially on each other (Figure 3.24) (Nurnus *et al.*, 2008). The growth of thin-film IV–VI /V_2–VI_3 SLs has been also reported.

A precondition for the epitaxial growth is an almost similar lattice constant. This is given for the IV–VI lattice constant in the 111 plane (PbSe: a = 4,330A; PbTe: a = 4,572A) and the V_2VI_3 lattice constant in the 001 plane (Bi_2Te_3 a = 4,384A; Bi_2Se_3 a = 4,130A; Sb_2Te_3 a = 4,250A).

The transfer of this thin-film concept to bulk materials requires further conditions. An extended miscibility gap is needed to obtain pure semiconductors and not a solid solution.

All these premises are fulfilled for PbTe/Sb_2Te_3 and PbTe/Bi_2Te_3 composites.

A lamellar SL-like microstructure of PbTe and Sb_2Te_3 with epitaxial-like interfaces was achieved by decomposing a metastable phase (Ikeda, 2007). The inter-lamellar spacing can be controlled through the recrystallizsation process. A spacing of 40 nm PbTe and 180 nm Sb_2Te_3 was obtained.

The reduction of the lattice thermal conductivity of PbTe/Sb_2Te_3 lamellar samples decreases with decreasing inter-lamellar spacing (Ikeda *et al.*, 2007). It was considered to be due to the coarsening of the microstructure. Nevertheless there are further challenges for improving the *ZT* of this material, because it is complicated to optimise the carrier concentration regarding optimum thermoelectric properties.

3.2.2.5 Summary and outlook

The state of art for nanoscale thermoelectrics in science and applications demonstrates clearly the potential of the nanoscale concept to increase significantly the material quality. The expectation to reduce thermal conductivity by enhanced phonon scattering is verified for different types of nanostructures. The technical ripeness was proved by the successful transfer of academic results of RTI into the Nextreme company.

It should be stated that up to now, noticeable hurdles have to be mastered on the way from lab to industrial mass production as far as energy conversion in the kW range, as expected for cars, is concerned.

One main challenge will be the temperature term stability of any of the nanoscale structures.

The dominant material in the field of thermoelectrics is stable against diffusion up to about $350°C$. But it was shown (Böttner, 2004) that significant diffusion will take place in nanoscale V–VI compositions at remarkable lower temperatures, even down to $\sim100°C$.

Improved efficiencies for nanoscale thermoelectric materials, like reported here, can also be achieved with bulk-like materials with 'built-in' nanostructures. One example of such a thermoelectric material is a certain high-temperature material called *skutterudites* after the Norwegian mineral deposit in Skutterud. In skutterudites, the elements cobalt and antimony form a more or less open crystal lattice that can accommodate heavy atoms as so-called rattlers in its voids, say, as a 'built-in nanostructure'. Such 'filled' skutterudites have comparatively low thermal conductivity but high electrical conductivity.

The examples are conclusive for all these advanced thermoelectric materials. Thus energy harvesting using thermoelectric is within tangible reach.

Bibliography

Böttner, H., Schubert, A., Kölbel, H., Gavrikov, A., Mahlke, A., and Nurnus, J. (2004). Formation of nanometerscale layers of V–VI (Bi_2Te_3-related) compounds based on amorphous prestages, Proc. 23rd Int. Conf. Thermoelelctrics (ICT), Adelaide, Australia, July 25–29, 2004, 114.

Dresselhaus, M. (1992). Proc. First Thermogenic Workshop, Ft. Belvoir; VA, Stu Hron, USA.

Ebling, D., Jacquot, A., Jägle, M., Böttner, H., Kühn, U., and Kirste, L. (2007). *Phys. Stat. Sol.* (RRL) 1(6), 238–240.

Gothard, Spowart, and Tritt. (2010). *Phys. Status Solidi* A, 207(1), 157–162.

Harman. (2000). MIT-LL, *J. Elec. Mat.*, 2000.

Harman, T. C., Taylor, P. J., Walsh, M. P., and LaForge, B. E. (2002). *Science*, 297, 2229–2232.

Harmann, T. C., Walsh, M. P., LaForge, B. E., and Turner, G. W. (2005). *J. Elec. Mat.*, 34(5), L19–L22.

He, Z. M., Stiewe, C., Platzek, D., Karpinski, G., Mueller, E., Li, S. H., Toprak, M., and Muhammed, M. (2007). *Nanotechnology*, 18, 235602.

Hicks, L. D. and Dresselhaus. (1993). M. St Thermoelectric figure of merit of a onedimensional conductor, *Phys. Rev.* B, 47, 16631–16634.

Hoang, K., Desai, K., and Mahanti, S. D. (2005). *Phys. Rev.* B, 72, 064102.

Hochbaum, A., Chen, R., Delgado, R. D., Liang, W., Garnet, E. C., Najarian, M., and Yang, P. (2008). Enhanced thermoelectric performance of rough silicon nanowires, *Nature*, 451, 163–167.

IIsu, K. F., Loo, S., Guo, F., Chen, W., Dyck, J. S., Uher, C., Hogan, T., Polychroniadis, E. K., and Kanatzidis, M. G. (2004). *Science*, 303, 818.

Ikeda, T. (2007). Self-assembled nanometer lamellae of thermoelectric PbTe and Sb_2Te_3 with epitaxy-like interfaces, *Chem. Mater.*, 19(4), 763–767.

Ikeda, T., Toberer, E., Ravi, V., Haile, S., and Snyder, J. (2007). Lattice thermal conductivity of self-assembled PbTe-Sb_2Te_3 composites with nanometer lamellae, Proc. Int. Conf. Thermoelectrics.

Kim, W., Zide, J., Gossard, A., Klenov, D., Stemmer, S., Shakouri, A., and Majumdar, A. (2006). *Phys. Rev. Lett.*, 96, 045901.

Li, D., Wu, Y., Kim, P., Shi, L., and Yang, P. (2003). *Appl. Phys. Lett.*, 83(14), 2934.

Nielsch, K., private communications.

Nurnus, J. and Böttner, H. (2000). Epitaxial growth and thermoelectric properties of Bi_2Te_3 based low dimensional structures, *Proc. Mater. Res. Soc.* Spring Meeting.

Nurnus, J., Künzel, C., Beyer, H., Lambrecht, A., Böttner, H., Meier, A., Blumers, M., Völklein, F., and Herres, N. (2000). Structural and thermoelectric properties of Bi_2Te_3 based layered structures, 19th Int. Conf. Thermoelectrics, Cardiff, U.K., 236–240.

Nurnus, J., Böttner, H., and Lambrecht, A. (2003). Thermoelectric micro devices: interplay of highly efficient thin film materials and technological compatibility, Proc. Inter. Conf. Thermoelectrics, Hérault - France, August 17–24, 2003.

Nurnus, J., Böttner, H., and Lambrecht, A. (2006). Nanoscale thermoelectrics, in D. M. Rowe (ed.), *Thermoelectrics Handbook: Macro to Nano-Structured Materials,* Taylor & Francis Boca Raton, 48.1–48.23.

Poudel, B., Hao, Q., Ma, Y., Lan, Y., Minnich, A., Yu, B., Yan, X., Wang, D., Muto, A., Vashaee, D., Chen, X., Liu, J., Dresselhaus, M. S., Chen, G., and Ren, Z. (2008). *Science*, 320, 634.

3.2.3 *Nanostructured Ceramic Membranes for Carbon Capture and Storage*

Dr. Martin Bram, Dr. Tim van Gestel,
Dr. Wilhelm Albert Meulenberg, and
Prof. Dr. Detlev Stöver
Forschungszentrum Jülich GmbH, Jülich, Germany

Membrane technology will play an important role in future CCS concepts, leading to low-CO_2-emission power generation. Today, gas separation membranes are developed based on all classes of materials, including polymers, metals, and ceramics. Their function is based on a number of different gas transport mechanisms, including molecular sieving, adsorption combined with surface diffusion, solution/diffusion mechanisms, ionic conduction, and mixed ionic and electronic conduction. In this work, the state of the art of microporous ceramic membranes is summarised. This kind of membrane must be structured on a nanometre scale to become suitable for CO_2 gas separation in future fossil power plant generations.

3.2.3.1 Requirements of membranes for gas separation in post- and pre-combustion power plants

For the next few decades, coal will continue to be one of the primary fuels for the growing global electricity production. To reduce the adversities of climate change caused by the emission of greenhouse gases, in particular CO_2, the development of low-emission or zero-emission fossil fuel technologies combined with CCS is a task that must be addressed. One important step in CCS is the separation of CO_2 from the flue gas or syngas. Today, two separation technologies appear most promising – chemical solvent technology and membrane technology.

Chemical solvent technology for CO_2 gas separation is already in commercial use in the chemical and natural gas industry and is therefore preferentially used in first-demonstration plants (Aaron, 2005; Herzog, 2001).

Solvent technology is based on an energy-consuming absorption/desorption process. It requires large amounts of toxic chemicals such as monoethanolamines (MEAs), which are difficult to recycle in terms of ecological aspects.

Hence, gas separation with membranes is thought to be an attractive alternative avoiding toxic chemicals, while also having the potential to reduce the energy demand for the separation process. If membrane technology is to be applied at the cold end of the flue gas stream in existing post-combustion power plants, some important issues have to be considered. The driving force for the separation process is the partial pressure difference between the feed flue gas and the permeate. Due to the low CO_2 content (approx. 14 mol. % after flue gas cleaning in hard-coal-fired power plants) and the ambient pressure of the flue gas, the driving force must be enhanced by extra measures. Therefore, different options exist: a) Increasing the partial pressure difference between feed and permeate using either a compressor on the feed side or a vacuum pump on the permeate side or combining both (Zhao, 2008); b) increasing the CO_2 content on the feed side by using oxyfuel combustion, which leads to a CO_2 fraction of approximately 80 mol. % in the flue gas (Yang, 2008). If using a multi-stage membrane system for post-combustion, the recirculation of the retentate back to the feed side is an effective measure, which may lead to the same goal (Zhao, 2009). Furthermore, if membrane technology is to be installed at the cold end of the flue gas stream, the possibility of acid corrosion of the membrane and housing materials has to be considered.

Pre-combustion capture processes appear to be an attractive alternative to post-combustion in terms of the integration of membrane technology. In this concept, the coal is gasified at high pressures (up to 60 bar) with pure oxygen or air, while a substoichiometric amount of water is added. If the gasification is done with pure oxygen, a mixture consisting mainly of CO and H_2 results. If air is used as the gasification medium, large amounts of N_2 must also be dealt with. CO reacts with water vapour in the water-gas shift reaction (WGS) to form CO_2 and H_2 in the presence of suitable catalysts.

Due to the high pressure of the gasification process and the high CO_2 concentration in the resulting gas mixture, pre-combustion provides a higher driving force for CO_2 separation processes than the conventional post-combustion process.

Detailed discussions on integrating membrane technology into the pre-combustion process can be found in literature (Marano, 2009; Grainger, 2008). Table 3.2 compares typical flue gas conditions and compositions, which the membranes are faced with in the case of post- and pre-combustion (Landes, 2009). The conditions and gas compositions may vary depending on the operation conditions of the power plant and the quality of the coal.

For post-combustion, membranes are preferred that allow CO_2 molecules to pass, while blocking other flue gas components. For pre-combustion, either CO_2-selective or H_2-selective membranes are suitable. Polymer as well as ceramic membranes are under development to fulfil these issues.

Table 3.2 Typical conditions and gas compositions for post- and pre-combustion (Landes, 2009)

Post-combustion		Pre-combustion	
Temperature	50…100°C	Temperature	200…400°C
Pressure	1,050 mbar	Pressure	up to 20 bar
Gas composition (mol. %)		Gas composition (mol. %)	
N_2	72	N_2	4
CO_2	13	CO_2	28
H_2O	11	H_2O	30
O_2	4	H_2	37
Trace gases (mg/m^3)		Trace gases (mg/m^3)	
SO_2	0.007	CO	0.6
NO_2	0.01	H_2S	0.002

Polymer membranes are already in use for gas separation on an industrial scale (Dortmundt, 1999; Nunes, 2006). Recently, promising selectivities for CO_2/N_2 were published, which make them especially attractive for the post-combustion process (Kim, 2004; Duan, 2006; Powell, 2006; Car, 2008). As an alternate to polymeric membranes, microporous ceramic membranes have been studied intensively for use in gas separation over the last few years.

They are expected to be advantageous if polymeric membranes do not achieve the required long-term stability in the presence of acid contaminants, steam, high temperatures, or high pressures. In this paper, the state of the art of these membranes is summarised.

3.2.3.2 Gas separation with microporous ceramic membranes

Gas separation with microporous membranes is based on selective diffusion of gas molecules through the membrane. Therefore, the size of the related gas molecules is an important issue for the development of gas separation membranes with a well-defined nanostructure. Table 3.3 summarises the kinetic diameters of the main gas molecules, which occur in post- and pre-combustion processes.

Table 3.3 Equilibrium length and width according to Pauling (1960) as well as kinetic diameters of the main gas molecules in post- and pre-combustion processes calculated from the minimum equlibrium diameter (Breck, 1974)

Molecule	Length (nm)	Width (nm)	Kinetic diameter (nm)
He	–	0.30	0.260
H_2O	0.39	0.315	0.265
H_2	0.31	0.24	0.289
CO_2	0.51	0.37	0.330
O_2	0.39	0.28	0.346
N_2	0.41	0.30	0.364
CO	0.42	0.37	0.376
NO	0.405	0.30	0.317
SO_2	0.528	0.40	0.360
H_2S	0.436	0.40	0.360

The kinetic diameter is defined as the closest distance between two molecules of the same species if they collide with zero initial kinetic energy (Breck, 1974). At this distance, the potential energy of interaction, which is described by the Lennard-Jones potential for spherical and non-polar molecules, respectively, and by the Stockmayer potential for polar molecules, approaches zero. For complex molecules, the minimum equilibrium molecule diameter according to Pauling (1960) is used to calculate the kinetic diameter.

To achieve selective gas transport through the membrane, the pore size must be in the range of the kinetic diameter of the molecules. According to the IUPAC notation, only microporous membranes (pore size < 2 nm, McCusker, 2001) are promising candidates. In practice, such membranes are supported by a graded multilayered ceramic structure, consisting of a macroporous substrate (pore size > 50 nm) and a mesoporous interlayer (pore size 2–50 nm). To adapt microporous membranes for gas separation, the different mechanisms of gas transport through microporous membranes must be considered. Figure 3.25 summarises the main transport mechanisms.

Viscous flow: If the pore size clearly exceeds the molecule size, all molecules may pass the membrane without any separation effect. Depending on the pore size, the tortuosity, and the pressure drop, laminar or turbulent gas flow through the membrane takes place.

Knudsen diffusion: If the pore size is slightly higher than the molecule size, the selectivity becomes proportional to the square

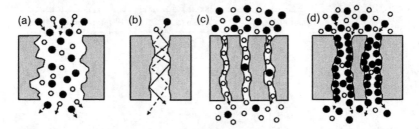

Figure 3.25 Principles of gas transport through ceramic membranes (Marano, 2009): (a) viscous flow, (b) Knudson diffusion, (c) molecular sieve, and (d) surface diffusion.

root of the molecular weight ratio of the gases, for example, 1.25 for CO_2/N_2 gas mixtures and 4.67 for H_2/CO_2 gas mixtures. Selectivities based on Knudsen diffusion are too low for gas separation on an industrial scale.

Surface diffusion: The situation of Knudsen diffusion changes clearly if one component of a gas mixture is preferentially adsorbed by the membrane material. If this adsorption occurs homogeneously all over the surface of the pore walls in the presence of a concentration gradient between the feed and permeate side, the surface diffusion of the adsorbed gas molecules leads to selectivity higher than Knudsen selectivity. Surface diffusion is the main mechanism for CO_2-selective membranes. Due to the reduction in adsorption capacity with increasing temperature, selectivity decreases with increasing temperature.

Molecular sieving (micropore diffusion): Smaller gas molecules may penetrate through the micropores of the membrane, while larger molecules are sieved out. The molecule transport through the membrane is a thermally activated process, with the rate increasing with temperature. Molecular sieving is the main mechanism for the separation of H_2 molecules from gas mixtures.

To achieve gas separation better than Knudsen diffusion with microporous ceramic membranes, the microstructure has to be adjusted on an atomistic scale, aligning the pore size of the membrane with the kinetic diameter of the respective molecule. In addition, doping of the membrane material to increase the adsorption capacity of the respective molecule is an effective measure, especially for the development of CO_2-selective membranes.

3.2.3.3 Membrane materials

Up to now, gas separation with ceramic membranes was demonstrated only with silica-based materials. Other ceramic materials like titania or zirconia, which should be promising candidates due to their high chemical resistance, are still lacking if gas separation based on a molecular diffusion process is the objective. Differences in chemical bonding of the membrane material appear to be responsible for this behaviour.

Microstructure of silica-based ceramics: The main building block of silica-based materials is the [SiO$_4$] tetrahedron, where the Si^{4+} cation is coordinated tetrahedrally by oxygen anions (Figure 3.26a). This structural assembly occurs preferentially if Si and O atoms with their ratio of the cation-anion-radii $r_{Si} : r_O = 0.30$ are combined (Scholze, 1988). The tetrahedra are linked among each other via oxygen bridges at the corners. In this configuration, the distance of adjacent Si^{4+} cations becomes maximal. The Si^{4+} cation can be easily exchanged by other cations like B^{3+} or Al^{3+}, but in this case alkali or earthalkali cations (Na$^+$, Li$^+$, Ca^{2+}) are required to balance the negative net charge. Depending on the material synthesis conditions and the amount of exchange cations, a glassy network or a well-ordered, crystalline network can be achieved (Figure 3.26b–d).

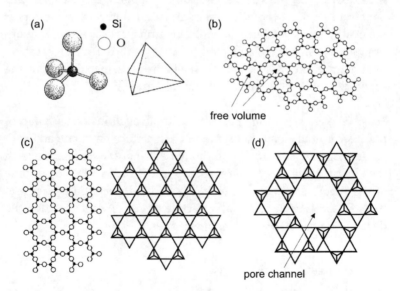

Figure 3.26 (a) [SiO$_4$] tetrahedron as the main building block of amorphous silica membranes, ball-and-stick model, solid tetrahedron as an analogon; (b) silica glass, ball-and-stick model; (c) quartz, ball-and-stick, as well as tetrahedron model, packing density 34 vol. %; (d) zeolite with well-defined pore channel, tetrahedron model.

The stable geometry of the tetrahedron with its well-defined atomic distances and the linkage of tetrahedra via the oxygen bridges lead to an atomic packing density of silica-based ceramics that is much lower than the atomic packing density of other ceramics or metals. For example, the maximum packing density of silica tetrahedra is achieved in a quartz crystal with 34 vol. % (Figure 3.26c). A further decrease in the packing density results when the tetrahedra are combined in a glassy network or a zeolite structure (Figure 3.26b,d). In both cases, a free volume on an atomistic scale results between the tetrahedra, which enables the adsorption and diffusion of gas molecules. Already in 1972, Shackelford *et al.*, (1972) demonstrated the diffusion of small He atoms and H_2 molecules in fused silica glasses. Furthermore, they found a reduction in the He diffusion coefficient when the silica glass contained network exchange ions like Na^+.

The most suitable class of materials for adsorption and diffusion of gas molecules are zeolites. Zeolites are usually crystalline alumi-nosilicates containing a regulary aligned pore network with very narrow pore channel diameters. They occur naturally as minerals, where more than 150 different structures are known (Breck, 1974), but for most practical applications, synthetic zeolites are preferred due to the lack of impurities and additional phases. Due to their unique porosity and the wide range of possible pore diameters, which are close to the molecule sizes of technically used gases, zeolites are applied in an industrial scale amongst others for the separation and removal of gases, mainly by adsorption/desorption processes using zeolite powder beds. Up to now, the application of zeolites for continous membrane processes has been limited by the difficulty of preparing zeolite membranes free of defects. In most cases, these membranes are synthesised by bringing the mesoporous substrate into contact with a zeolite seed solution. Afterwards, the zeolite layer nucleates and grows on the substrate. Caused by this preparation method, inter-crystalline pores can be left between the well-synthesised zeolite crystals, which limit the separation efficiency.

As an alternative, amorphous silica membranes have been investigated intensively over the last few years because of their ease and cost-effective preparation by sol-gel and CVD methods. In both cases, membranes with a thickness in the range of 50–200 nm could be prepared on multilayered substrates with a mesoporous interlayer. As expected from earlier results on fused silica glasses, the microstructure of amorphous silica membranes was found to be preferentially suitable for the diffusion of small molecules like He or H_2, while coarser molecules like CO_2 or CH_4 were reliably blocked. Due to the reduction in thickness, a clear increase in permeability was found compared with fused glasses. To also allow the adsorption and diffusion of coarser molecules through the glassy silica network, the application of templates, which were added temporarily during the sol-gel process, is a new and promising approach. When the templates are removed by a thermal treatment, a well-defined pore network results. If the pore diameter is in the range of the respective gas molecule, adsorption of respective molecules similar to zeolite membranes is expected. If the pore size exceeds the molecule size, subsequent doping of the pore walls by ion exchange or organic ligands, which enhances the affinity to the respective molecule, may also induce a gas separation based on a surface diffusion process.

Microstructure of nanostructured titania and zirconia: Due to the larger cation radii, it is not possible to prepare glassy microstructures or structures similar to zeolites from pure titania or zirconia (Scholze, 1988). Nevertheless, it is possible to prepare membranes from pure TiO_2 and ZrO_2 by sol-gel technology with a thickness less than 100 nm (Puhlfürß, 2000; Sekulic, 2004; vanGestel, 2008). It is assumed that the membrane consists of TiO_2 and ZrO_2 seed crystals (Figure 3.27), which are still X-ray amorphous due to their low crystallite size. These seed crystals were already formed in the sol by partial hydrolysis of the precursor and subsequent condensation. At the boundaries of the nanosized crystals, the lower degree of order of the atoms leads to an effective network of pores, which can be used, for example, for ultra- and nano-filtration, where pore sizes down to 1 nm are required. However, up to now, this kind of membranes has failed when gas separation was the aim. Another difficulty associated with these membranes is maintaining the small crystallite and pore size when the temperature is increased. In the temperature range of 300–500°C, grain growth of the nanosized crystallites begins. This is expected to change the performance of the membranes drastically.

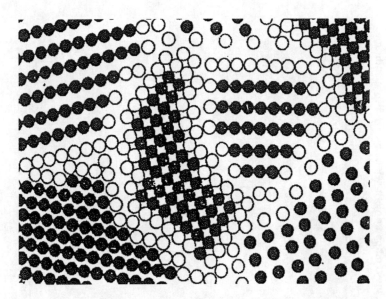

Figure 3.27 Schematic sketch of a nanocrystalline material (Birringer, 1984).

In the literature, some information can be found on the synthesis of TiO_2-based glasses by the addition of 20–60 mol. % K_2O, Na_2O, or BaO (Scholze, 1988). It is still unknown if these glasses are suitable for gas separation. In silica-based glasses, Si^{4+} cations can be exchanged with Ti^{4+} or Zr^{4+} cations up to a limited amount. This is known to enhance the chemical resistance of the glasses, especially under basic conditions. This approach is already used for gas separation membranes (Ohshima, 1998; Yoshida, 2001).

3.2.3.4 Performance of microporous ceramic membranes

In this overview, silica-based membranes with special properties/compositions are preferrentially considered, rather than giving a comprehensive summary of a large number of more or less identical membranes.

Silica-based membranes made by sol-gel and CVD techniques: Table 3.4 summarises a few selected silica membranes prepared by sol-gel technology and CVD. As discussed before, the glassy structure of the silica-based membranes is preferrentially suitable for the

Table 3.4 Properties of selected sol-gel SiO_2 and doped SiO_2 membranes compared with selected CVD SiO_2 membranes (permeabilities in mol/m²·s·Pa)

Reference	Membrane material	Measuring conditions	H_2 permeability	CO_2 permeability	N_2 permeability	H_2/CO_2 selectivity	CO_2/N_2 selectivity
DeVos, 1998	SiO_2 sol-gel	Single gas 200°C	$3.9\text{-}4.1 \cdot 10^{-7}$	$0.6\text{-}1.6 \cdot 10^{-8}$	< DL	37-66	–
Tsai, 2000	SiO_2 sol-gel	Single gas 80°C	$3.3 \cdot 10^{-7}$ *	$6.7 \cdot 10^{-8}$ *	$1.0 \cdot 10^{-9}$ *	5	60
Battersby, 2009	SiO_2 with Co sol-gel	Single gas 100°C	$5 \cdot 10^{-10}$	–	–	70	–
		250°C	$6 \cdot 10^{-9}$	–	–	1,000	–
Boffa, 2008	SiO_2 with Nb_2O_5 sol-gel	Single gas 200°C	$3.9 \cdot 10^{-8}$	$8.5 \cdot 10^{-10}$	$6.2 \cdot 10^{-9}$	46	13
Yoshida, 2001	SiO_2 with ZrO_2 sol-gel	Single gas 300°C	$1.7 \cdot 10^{-7}$ *	$8.8 \cdot 10^{-9}$ *	$4.0 \cdot 10^{-9}$ *	20	2.2
Hauler, 2009	SiO_2 sol-gel	Single gas 25°C	$5.5 \cdot 10^{-8}$	$5.1 \cdot 10^{-10}$	< DL	107	–
		100°C	$1.1 \cdot 10^{-7}$	$3.0 \cdot 10^{-10}$	< DL	372	–
		200°C	$1.6 \cdot 10^{-7}$	$3.8 \cdot 10^{-10}$	< DL	410	–
Hauler, 2009	SiO_2 with Co_3O_4 sol-gel	Single gas 25°C	$1.3 \cdot 10^{-8}$	< DL	< DL	> 1,000	–
		100°C	$2.8 \cdot 10^{-8}$	$4.8 \cdot 10^{-10}$	< DL	58	–
		200°C	$6.6 \cdot 10^{-8}$	$8.4 \cdot 10^{-10}$	< DL	78	–

Hauler, 2009	SiO$_2$ with ZrO$_2$ sol-gel	Single gas 200°C	$8.5 \cdot 10^{-8}$	$5.7 \cdot 10^{-9}$	$1.2 \cdot 10^{-9}$	15	5
Oyama, 2004	SiO$_2$ CVD	Single gas 600°C	$1.8 \cdot 10^{-7}$	$8.1 \cdot 10^{-11}$	–	2,200	–
Gu, 2008	SiO$_2$ with Al$_2$O$_3$ CVD	Single gas 600°C	$1.6 \cdot 10^{-7}$	$2.5 \cdot 10^{-10}$	–	590	–
Sea, 1997	SiO$_2$ CVD	Single gas 200°C	$2.2 \cdot 10^{-8}$	$3.3 \cdot 10^{-10}$	$5.0 \cdot 10^{-10}$	67	4.2

*Permeability converted by the author; DL = Detection limit.

separation of H_2 from CO_2. Additionally, the latest results achieved in our laboratory have been added (Hauler, 2009). Generally, doping with metals (Co, Ni) or suitable oxides (NiO, Co_3O_4, TiO_2, ZrO_2) has been become increasingly interesting in the last few years in terms of enhancing the limited hydrothermal stability of pure SiO_2. Even if the mechansims of stabilisation have not been fully understood up to now, an improved stability against water-vapour-containing atmospheres has been demonstrated.

Templated SiO_2 membranes: During the sol-gel process, a template was incorporated into the SiO_2 network, which was removed by a thermal treatment after membrane synthesis. As a result, an ordered pore structure was achieved in the glassy network. Templates, which are frequently used, are cationic surfactants (e.g., tetra-propyl-ammonium-bromide [TPAB]) or anionic surfactants (e.g., Brij = poly-ethylene-cetyl, lauryl, or stearyl ether). Table 3.5 gives an overview of a few selected membranes. Particularly for separation of larger gas molecules, such as CO_2 from N_2, these membranes can be a promising alternative to zeolite membranes due to the adjusted pore radius. However, H_2/CO_2 separation was also demonstrated with such membranes.

Zeolite membranes: A number of results on zeolite membranes have been reported, which yield a significant separation for CO_2, especially when the measuring temperature was below or near room temperature. This result is explained on the basis of high CO_2 adsorption capacity of the zeolite membrane material and the occurrence of a surface diffusion mechanism. At low temperatures, CO_2 adsorption may even block the diffusion of smaller H_2 molecules through the membrane, leading to CO_2/H_2 gas separation similar to polymer membranes (Hong, 2008). The CO_2 adsorption capacity and CO_2/N_2 selectivity were further improved when basic cations were incorporated into the zeolithic structure by immersing the zeolite membrane in a solution of alkali or earth-alkali salts followed by a thermal treatment. Table 3.6 gives an overview of a few selected zeolite membranes as reported in the literature.

Table 3.5   Properties of selected SiO_2 with organic templates (permeabilities in mol/m²·s·Pa)

Reference	Membrane material	Measuring conditions	H_2 permeability	CO_2 permeability	N_2 permeability	H_2/CO_2 selectivity	CO_2/N_2 selectivity
Kim, 2001	SiO_2 with 10% template sol-gel	Single gas					
		100°C	$2.4 \cdot 10^{-7}$	$2.0 \cdot 10^{-8}$	$3.1 \cdot 10^{-9}$	12	6
		200°C	$4.0 \cdot 10^{-7}$	$1.8 \cdot 10^{-8}$	$4.5 \cdot 10^{-9}$	22	4
		300°C	$6.8 \cdot 10^{-7}$	$2.0 \cdot 10^{-8}$	$2.3 \cdot 10^{-8}$	34	0.8
		Gas mix H_2/CO_2					
		100°C	$1.1 \cdot 10^{-8}$	$9.7 \cdot 10^{-9}$	–	11	–
		200°C	$1.8 \cdot 10^{-8}$	$9.1 \cdot 10^{-9}$	–	20	–
		300°C	$2.9 \cdot 10^{-8}$	$8.2 \cdot 10^{-9}$	–	36	–
		Gas mix CO_2/N_2					
		100°C	–	$2.0 \cdot 10^{-8}$	$1.5 \cdot 10^{-9}$	–	13
		200°C	–	$2.0 \cdot 10^{-8}$	$3.3 \cdot 10^{-9}$	–	16
		300°C	–	$2.1 \cdot 10^{-8}$	$1.3 \cdot 10^{-8}$	–	1.6
Ohshima, 1998	SiO_2 with ZrO_2 and organic template, sol-gel	Gas mix CO_2/N_2					
		25°C	–	$8 \cdot 10^{-8}$	–	–	60
		100°C	–	$2 \cdot 10^{-7}$	–	–	20
		200°C	–	$4 \cdot 10^{-7}$	–	–	5

Table 3.6 Properties of selected zeolite membranes (permeabilities in $mol/m^2 \cdot s \cdot Pa$)

Reference	Membrane material	Measuring conditions	H_2 permeability	CO_2 permeability	N_2 permeability	CO_2/H_2 selectivity	CO_2/N_2 selectivity
Hong, 2008	Zeolite SAPO-34	Single gas 37°C	$1-2 \cdot 10^{-8}$	$3.5-4 \cdot 10^{-8}$	–	$2-3.5$	–
		Gas mix −20°C	$2.0 \cdot 10^{-9}$	$2.0 \cdot 10^{-8}$	–	10	–
		37°C	$2.5 \cdot 10^{-8}$	$3.5 \cdot 10^{-8}$	–	1.7	–
Sebastian, 2007	MFI-zeolite ZSM-5	Gas mix CO_2/N_2 25°C	–	$2.7 \cdot 10^{-7}$	–	–	13
Bonhomme, 2003	MFI-zeolite ZSM-5	Single gas 25°C	$1.9-2.9 \cdot 10^{-7}$	$2.9-5.9 \cdot 10^{-7}$	$1.1-1.6 \cdot 10^{-7}$	$1.5-2$	$3-4$
Kusakabe, 1998	NaY zeolite ionex-changed with	Gas mix CO_2/N_2 40°C pure NaY	–	$1.5 \cdot 10^{-6}$	–	–	19.6
	Li	with Li	–	$2.0 \cdot 10^{-6}$	–	–	3.5
	K	with K	–	$1.5 \cdot 10^{-6}$	–	–	30.3
	Mg	with Mg	–	$3.0 \cdot 10^{-7}$	–	–	20.5
	Ca	with Ca	–	$5.0 \cdot 10^{-7}$	–	–	11.8
	Ba	with Ba	–	$8.0 \cdot 10^{-7}$	–	–	23.7
Hasegawa, 2002	Y-type zeolite ionex-changed with K,Rb,Cs	Gas mix CO_2/N_2 35°C	–	$1.0 \cdot 10^{-6}$	–	–	$30-150$

3.2.3.5 Summary and conclusion

Microporous ceramic membranes are thought to be promising candidates for the CO_2 gas separation in future fossil power plants. Up to now, gas separation has only been demonstrated with two classes of silica-based membranes, glassy silica membranes and zeolites. Both classes are based on a $[SiO_4]$ tertrahedron as the main building block. A glassy arrangement of these tetrahedra was found to be suitable for a thermally activated diffusion of small H_2 molecules through the low-ordered silica network, while coarse CO_2 molecules were blocked. This effect can be used, for example, for H_2/CO_2 separation in a pre-combustion process. Contrary to this, the tetrahedra in zeolites are arranged in crystalline structures, including well-defined pore channels with pore diameters in the range of the respective molecules. With zeolite membranes, CO_2 separation from binary CO_2/N_2 gas mixtures was demonstrated, which is particularly attractive for post-combustion applications. In this case, gas separation is based on the preferential adsorption of CO_2 molecules by the zeolite material followed by a CO_2 surface diffusion along the pore channel walls. Due to the difficulties in preparing zeolite membranes avoiding coarser inter-crystalline pores, which limit the membrane selectivity, current research focuses on developing more effective, alternative synthesis methods and improving the separation process by innovative surface modification. It is important to mention that gas separation with microporous ceramic membranes has been demonstrated mainly on a laboratory scale so far using preferrentially single gases or binary gas mixtures. Therefore, tests under real conditions, for example, directly in power plants, are required to demonstrate the long-term stability of membrane materials and membrane performance in contact with moisture, ash particles, trace gases, and other influencing factors.

Bibliography

Aaron, D. and Tsouris, C. (2005). Separation of CO_2 from flue gas: a review, *Sep. Sci. Technol.*, 40, 321–348.

Battersby, S., Tasaki, T., Smart, S., Ladewig, B., Liu, S., Duke, M. C., Rudolph, V., and Diniz da Costa, J. C. (2009). Performance of cobalt silica membranes in gas mixture separation, *J. Membr. Sci.*, 329, 91–98.

Birringer, R., Gleiter, H., Klein, H. P., and Marquardt, P. (1984). Nanocrstalline materials – An approach to a novel solid structure with gas-like disorder? *Phys. Lett.*, 102A, 365–369.

Boffa, V., Blank, D. H. A., and ten Elshof, J. E. (2008). Hydrothermal stability of microporous silica and niobia–silica membranes, *J. Membr. Sci.*, 319, 256–263.

Bonhomme, F., Welk, M. E., and Nenoff, T. M. (2003). CO_2 selectivity and lifetimes of high silica ZSM-5 membranes, *Microp. Mesop. Mat.*, 66, 181–188.

Breck, D. W. (1974). Zeolite Molecular Sieves: Structure, Chemistry, and Use, John Wiley & Sons, New York, London, Sydney, Tokyo.

Car, A., Stropnik, C., Yave, W., and Peinemann, K. V. (2008). Tailor made polymeric membranes based on segmented block copolymers for CO_2 separation, *Adv. Funct. Mat.*, 18, 2815–2823.

De Vos, R. M. and Verweij, H. (1998). Improved performance of silica membranes for gas separation, *J. Membr. Sci.*, 143, 37–51.

Dortmundt, D. and Doshi, K. (1999). Recent Developments in CO_2 Removal Membrane Technology, UOP LLC, Des Plaines, IL, USA.

Duan, S., Kouketsu, T., Kazama, S., and Yamada, K. (2006). Development of PAMAM dendrimer composite membranes for CO_2 separation, *J. Membr. Sci.*, 283, 2–6.

Grainger, D. and Hägg, M. B. (2008). Techno-economic evaluation of a PVAm CO_2-selective membrane in a IGCC power plant with CO_2 capture, *Fuel*, 87, 14–24.

Gu, Y., Harcarlioglu, P., and Oyama, S. T. (2008). Hydrothermally stable silica-alumina composite membranes for hydrogen separation, *J. Membr. Sci.*, 310, 28–37.

Hasegawa, Y., Watanabe, K., Kusakabe, K., and Morooka, S. (2002). Influence of alkali cations on permeation properties of Y-type zeolite membranes, *J. Membr. Sci.*, 208, 415–418.

Hauler, F. and van Gestel, T. (2009). Internal report, Forschungszentrum Juelich.

Herzog, H. (2001). What future for carbon capture and sequestration? *Environ. Sci. Technol.*, 35, 148–153.

Hong, M., Li, S., Falconer, J. L., and Noble, R. D. (2008). Hydrogen purification using a SAPO-34 membrane, *J. Membr. Sci.*, 307, 277–283.

Kim, T. J., Li, B., and Hägg, M. B. (2004). Novel fixed-site-carrier polyvinylamine membrane for carbon dioxide capture, *J. Polym. Sci., Part* B: Polym. Phys., 42, 4326–4336.

Kim, Y. S., Kusakabe, K., Morooka, S., and Yang, S. M. (2001). Preparation of microporous silica membranes for gas separation, *Korean J. Chem. Eng.*, 18(1), 106–112.

Kusakabe, K., Kuroda, T., and Morooka, S. (1998). Separation of carbon dioxide from nitrogen using ion-exchanged faujasite-type zeolite membranes formed on a porous support tube, *J. Membr. Sci.*, 148, 13–23.

Landes, H. (2009). Personal communication.

Marano, J. J. and Ciferino, J. P. (2009). Integration of gas separation membranes with IGCC, *Energy Procedia*, 1, 361–368.

McCusker, L. B., Liebau, F., and Engelhardt, G. (2001). Nomenclature of structural and compositional characteristics of ordered microporous and mesoporous materials with inorganic hosts, *Pure Appl. Chem.*, 73, 381–394.

Nunes, S. P. and Peinemann, K. V. (2006). Membrane Technology in the Chemical Industry, Wiley-VCH Verlag GmbH, *Germany*, 53–150.

Ohshima, Y., Seki, Y., and Maruyama, H. (1998). Permselectivity of SiO_2 membrane doped with both ZrO_2 and organic ligand, in S. Nakao (ed.), Proc. 5th Int. Conf. Inorg. Membr., *Nagoya*, 668–671.

Oyama, S. T., Lee, D., Hacarlioglu, P., and Saraf, R. F. (2004). Theory of hydrogen permeability in nonporous silica membranes, *J. Membr. Sci.*, 244, 45–53.

Pauling, L. (1960). Nature of Chemical Bond, Cornell University Press, Ithaca, NY, USA.

Powell, C. E. and Qiao, G. G. (2006). Polymeric CO_2/N_2 gas separation membranes for the capture of carbon dioxide from power plant flue gases, *J. Membr. Sci.*, 279, 1–49.

Puhlfürß, P., Voigt, A., Weber, R., and Morbé, M. (2000). Microporous TiO_2 membranes with a cut-off < 500 Da, *J. Membr. Sci.*, 174, 123–133.

Sea, B. K., Kusakabe, K., and Morooka, S. (1997). Pore size control and gas permeation kinetics of silica membranes by pyrolysis of phenyl-substituted ethoxysilanes with cross-flow through a porous support wall, *J. Membr. Sci.*, 130, 41–52.

Scholze, H. (1988). Glas – Natur, Struktur und Eigenschaften, Springer-Verlag, Berlin, Heidelberg, New York, London, Paris, Tokyo.

Sebastian, V., Kumakiri, I., Bredesen, R., and Menendez, M. (2007). Zeolite membrane for CO_2 removal: operating at high pressure, *J. Membr. Sci.*, 292, 92–97.

Sekulic, J., ten Elshof J. E., and Blank, D. H. A. (2004). A microporous titania membrane for nanofiltration and pervaporation, *Adv. Mat.*, 72, 49–57.

Shakelford, J. F. Studt, P. L., and Fulrath, R. M. (1972). Solubility of gases in glass. II. He, Ne, and H_2 in fused silica, *J. Appl. Phys.*, 43, 1619–1626.

Tsai, C. Y., Tam, S. Y., Lu, Y., and Brinker, C. J. (2000). Dual-layer asymmetric microporous silica membranes, *J. Membr. Sci.*, 169, 255–268.

Van Gestel, T., Meulenberg, W. A., Bram, M., and Buchkremer, H. P. (2008). Manufacturing of new nanostructured ceramic-metallic composite nanofiltration and gas separation membranes consisting of ZrO_2, Al_2O_3, TiO_2 and stainless steel, *Solid State Ionics*, 179, 1360–1366.

Yang, H., Xu, Z., Fan, M., Gupta, R., Slimane, R. B., Bland, A. E., and Wright, I. (2008). Progress in carbon dioxide separation and capture: a review, *J. Environ. Sci.*, 20, 14–27.

Yoshida, K. (2001). Hydrothermal stability and performance of silica-zirconia membranes for hydrogen separation in hydrothermal conditions, *J. Chem. Eng. Jpn.*, 34, 523–530.

Zhao, L., Riensche, E., Menzer, R., Blum, L., and Stolten, D. (2008). A parametric study of CO_2/N_2 gas separation membrane processes for post-combustion capture, *J. Membr. Sci.*, 325, 284–294.

Zhao, L., Menzer, R., Riensche, E., Blum, L., and Stolten, D. (2009). Concepts and investment cost analyses of multi-stage membrane systems used in post-combustion processes, *Energy Procedia*, 1, 269–278.

3.3 Energy Storage and Distribution

3.3.1 *Materials for Energy Storage*

Dr. Wiebke Lohstroh
Forschungsneutronenquelle Heinz Maier-Leibnitz (FRM II), Technische Universität München, Germany

The storage of energy is one of the most challenging tasks of today. In all likelihood, the use of fossil fuels will have to be curtailed in future, both due to economic reasons as supply is limited and due to ecologic reasons to cut CO_2 emissions. Thus it is mandatory to develop and utilise other means of energy generation and storage. The increasing percentage of renewable energy sources in the energy mix, such as wind power or PV panels, subjects the supply of electric energy to major fluctuations. Since supply and demand will not always be in balance, intermittent energy storage is a mandatory. These storage systems should be optimised for energy density and efficiency as well as cost and life time. In addition, energy storage systems are required to stabilise the grid buffering peaks (or dips) in the energy supply on very short time scales. Here optimisation for response time and rate capabilities is required. The currently investigated or applied techniques can be differentiated into several groups: (i) mechanical storage such as water pump storage, flywheel mechanical storage, or compressed air storage in underground caverns, (ii) electrical storage in supercapacitors or superconducting coils, or (iii) electrochemical storage, for example, in batteries or hydrogen generation in an electrolyser.

All these techniques have specific strengths and weaknesses, and no unique solution will be suited for all different applications.

Another challenging task is the replacement of fossil fuels in mobile application such as cars, buses, or heavy-duty vehicles. Here, the weight and the volume of the storage system are of major importance. The currently used petrol has a high gravimetric and volumetric energy density, and due to its liquid form, on board storage and refueling are easily realised at low cost.

For mobile applications, mainly chemical energy storage systems are investigated, especially in the form of hydrogen-powering a fuel cell, or electrochemical storage in form of a secondary battery. The energy content of various systems is schematically presented in Figure 3.28.

In today's view, battery-powered vehicles are well suited for inner-city mobility with its short distances and frequent start/stop cycles. On the other side, long-distance or inter-city travel is not fore-seeable with the energy densities of the currently available battery techniques and other options have to be considered, such as hybrid cars or the use of range extenders. For CO_2-emission-free vehicles, fuel cell cars running on hydrogen might be more advantageous.

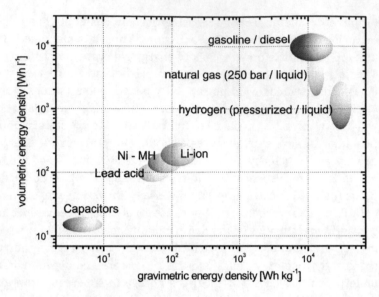

Figure 3.28 Schematic representation of the gravimetric and volumetric energy density of various systems. See also Colour Insert.

In batteries, the storage of energy and the regeneration of electrical power are combined in a single system, that is, the choice of cell chemistry fundamentally determines the maximum energy density of the battery which then has to be developed for the intended application.

For fuel cells the storage of hydrogen and its conversion into electric power is separated into different parts, that is, in the storage part and the power-generating fuel cell. At first glance this appears to be a more complicated solution; however, it also offers more degrees of freedom to tackle specific problems. Nevertheless, the requirements for both battery-driven cars and fuel cell cars are very similar: The amount of stored energy in terms of weight and volume should allow a certain driving range and offer reasonable power density. In addition, refueling should be fast and easy to handle and the entire system should be reliable, safe during operation and production, be environmentally friendly, be energy efficient, and come at a low cost. The further development of materials for energy storage will be a necessary prerequisite in achieving these goals. Of crucial importance for electrochemical means of energy storage will be the ability to control and guide the underlying chemical reactions. Nanotechnological methods are well suited to approach these tasks, and great advances in the performance of energy storage systems can be expected in the following years.

In the following, an overview will be presented of the currently investigated materials for energy storage both for hydrogen storage as well as for batteries.

3.3.1.1 Materials for hydrogen storage

Hydrogen as an energy carrier is attractive due to its very high energy density per mass, that is, 33 kWh/kg in comparison with 13.9 kWh/kg for natural gas or 12.7 kWh/kg for petrol. Moreover, only water is produced upon combustion either in internal combustion engines or in fuel cells. Today's proton exchange membrane (PEM) fuel cells have reached a high level of maturity and their reliability in vehicular applications has been demonstrated. One major issue is still the high cost mainly due to the high amount of platinum needed to catalyse the reaction.

Although hydrogen has a high gravimetric energy density, it is a gas with a low volumetric energy density. Currently, storage is realised either in compressed form (in tanks up to 70 MPa) or as cryogenic liquid at 21 K.

Both methods are considered mature techniques, and it appears as if the 70 MPa pressurised tank will be the choice for the first vehicles entering small-scale series production. Although these storage techniques are well developed, their drawback is the considerable amount of energy spent for compression or cooling (up to one-third of the lower heating value or hydrogen) (Eberle *et al.*, 2009). In addition, the volumetric storage capacity is still low and the storage systems are usually quite bulky.

Another possibility is hydrogen storage in a solid-state storage tank in which hydrogen is bound to a host material. A variety of materials takes up hydrogen either as a surface-bound molecule (i.e., physical storage) or as an atom chemically bound in the host lattice (see Figure 3.29).

The density of atomically bound hydrogen in a host material can easily exceed that of liquid hydrogen, and the volumetric storage

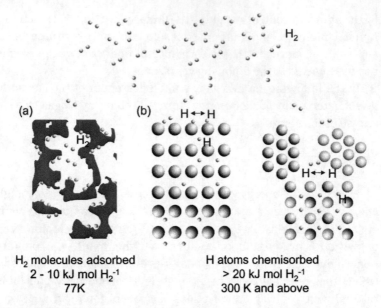

H$_2$ molecules adsorbed
2 - 10 kJ mol H$_2^{-1}$
77K

H atoms chemisorbed
> 20 kJ mol H$_2^{-1}$
300 K and above

Figure 3.29 Solid-state hydrogen storage: (a) adsorption of hydrogen molecules in high-surface-area hosts at low temperatures (physisorption) and (b) atomic hydrogen chemically in a host lattice, either on interstitials (left panel) or in a stoichiometric compound (right panel).

density becomes very high although at the expense of the additional weight of the storage material. Typical binding energies for physical storage (mainly van der Waals interactions) are in the order of $\sim$2–10 kJ mol^{-1} H$_2$, and the systems operate at liquid nitrogen temperature (77 K). The relevant energies for chemically bound hydrogen are in the range of >20 kJ mol^{-1} H$_2$, and the operating temperatures are equal or above room temperature.

Key characteristics for both storage methods are the maximum hydrogen capacity, its exchange rate, and the operating temperature; all these parameters crucially depend on the microstructure and processing of the active components.

3.3.1.2 Physiorption materials

Materials with high surface areas and high porosities are investigated for adsorption hydrogen storage; among them are high-surface carbon materials, zeolites, metal organic frameworks

(MOFs), and porous polymers. The hydrogen storage capacity generally correlates with the apparent surface area (Langmuir or BET surface area measured by N_2 adsorption) and the available free pore volume of the host material (Panella *et al.*, 2006; Wong-Foy *et al.*, 2006; Langmi *et al.*, 2003). At low pressures (0.1 MPa) and temperatures ($\sim$77 K), a linear relationship between the gravimetric storage capacity and the surface area has been reported; for carbon the proportionality constant is $\sim$1.3 wt% 10^{-3}g m^{-2} (Thomas, 2007; Thomas, 2009). From this relationship and the free surface of a graphene sheet ($\sim$2,630 m^2 g^{-1}), the limiting capacity of carbon-based materials can be estimated to be around 3.4 wt% H_2.

Higher surface areas can be obtained in metal organic frameworks ($>$3,000 m^2 g^{-1}), and correspondingly higher hydrogen uptake has been claimed (e.g., MOF 177, $>$ 4,500 m^{2}g^{-1} and 7.5 wt% H_2) (Rowswell *et al.*, 2004). However, hydrogen uptake in MOFs does not necessarily saturate at ambient pressures, and high-pressure experiments are needed to evaluate the full potential.

It has been suggested that the hydrogen density in porous materials is limited by the density of liquid hydrogen (Zuttel *et al.*, 2002). Small pores below 1 nm seem to be beneficial to reach that limiting value as hydrogen binds more strongly in cavities with smaller diameters (Hirscher *et al.*, 2010). Thus the challenge is the design and development of materials providing large surface areas in combination with small pore diameters.

3.3.1.3 Chemisorption materials

In chemisorption materials hydrogen atoms are bound to a host lattice. The binding energies are generally higher than for physisorption materials, and the operating temperatures are room temperature or above. The underlying mechanism involves several distinct steps: hydrogen molecules are adsorbed at the surface, then they are split into atomic hydrogen, and finally the hydrogen atoms diffuse into the bulk where they are bound on specific sites. Upon desorption, the entire process has to be reversed. Thermodynamically, temperature and pressure for hydrogen uptake and release follow the van't Hoff equation:

$$\ln p/p_0 = \Delta H/RT - \Delta S/T \qquad (3.6)$$

where ΔH and ΔS are the enthalpy and entropy change of the reaction. The equilibrium pressure of 0.1 MPa hydrogen pressure is reached at $T_{eq} = \Delta H / \Delta S$. This yields a desired value for the enthalpy change of $\Delta H \sim -40$ kJ/mol H_2 for hydrogen absorption, that is, 0.1 MPa equilibrium pressure at room temperature (assuming $\Delta S = 130$ J/ mol H_2 K, the entropy change of 1 mol H_2 gas to the dissolved state at standard conditions).

However, real systems rarely exhibit the thermodynamic decomposition temperature as kinetic barriers shift the actual temperatures to much higher values. This is especially the case for alkaline earth metal hydrides, for example, MgH_2, or complex hydrides discussed below.

Traditionally, transition metal hydrides such as $LaNi_5H_x$ were investigated as hydrogen storage materials. Here, hydrogen atoms reside on interstitials of the host lattice and the materials remain metallic over a wide range of hydrogen concentrations.

The hydrogen atoms exhibit rather high diffusion coefficients in the order of $D \geq 10^{-11}$ m^2/s at room temperature. However, their gravimetric storage capacity is at maximum ~ 2 wt% H_2 due to the heavy host material. That is why storage systems solely based on intermetallic compounds are not very attractive for mobile applications. However, Toyota presented a study of a hybrid tank combining a high-pressure vessel (35 MPa) and a Ti-Cr-Mn hydrogen storage alloy. The combined system has a capacity (at 350 bar H_2) of 7.3 kg H_2 in 180 L and a weight of 420 kg (Mori, 2005).

In order to enhance the gravimetric storage capacity, the focus has shifted to either light alkaline (earth) hydrides (e.g., LiH or MgH_2) or complex hydrides based on light metals (such as $LiBH_4$). These materials offer much higher hydrogen content, but reaction rates and reversibility are generally poor. The challenges are multifaceted and naturally vary for the different systems, but two common main focal points are:

- Enhancement of hydrogen exchange reaction rates (for materials with suitable thermodynamic properties), that is, lowering kinetically determined operating temperatures, enhancing reversibility, and prolonging cycle life, and
- Tailoring thermodynamic properties for materials with high hydrogen content but unsuitable thermodynamics.

Clearly, nanostructured designs with small particle sizes and a high density of reaction sites are well suited to address the first issue. Similarly, thermodynamic properties can be adjusted in nanocomposites comprising reaction-based systems using small particles a few nanometres in size only. In all cases, the crucial parameters are the control of particles sizes (preferentially in the submicron range or smaller), the preparation of composites with enhanced functionality, or the development of new additives increasing the exchange reaction rates.

3.3.1.3.1 Salt-like hydrides Hydrides based on the light alkaline or alkaline earth metals such as LiH, NaH and, MgH_2 show high gravimetric hydrogen storage capacities of 12.6, 4.2, and 7.6 wt% H_2. The stability of these compounds is quite high with decomposition temperatures of $\sim$720°C, 425°C, and 330°C for LiH, NaH, and MgH_2 (at 0.1 MPa H_2), respectively (Chase, 1998). In addition, these materials are stoichiometric compounds and the hydrogen rich phases are wide-band-gap insulators with poor electrical and thermal conductivity. Hydrogen diffusion inside the hydride phase is in lower orders of magnitudes compared with metal hydrides such as $LaNi_5H_x$, and consequently, once the outer surface of a particle is hydrogenated, further hydrogen uptake is severely hampered.

Among the salt-like binary hydrides, MgH_2 has been intensely studied for storage applications (Aguey-Zinsou and Ares-Fernandez, 2010) due to its storage capacity, abundance, and low cost.

The kinetic performance is significantly improved by the addition of catalysts (Dornheim *et al.*, 2006; Dornheim *et al.*, 2007; Barkhordarian *et al.*, 2006; Hout *et al.*, 2001; Gross *et al.*, 1998; Gross *et al.*, 1997). The best results have been obtained with transition metal oxides such as Nb_2O_5, and high reaction rates have been demonstrated over hundreds of sorption cycles. Moreover, the reduction of particle sizes by high energy ball milling is beneficial: submicron-sized particles ($\sim$500 nm) release and reabsorb hydrogen in a few hundred seconds, while for $\sim$5 μm particles, the same reaction takes hours for completion. The role of the additive is still under discussion: apparently it facilitates the splitting and recombination of hydrogen molecules, but noble metal catalysts such as palladium typically used for this task do not perform better compared with transition metal oxides. It has also been suggested that the stabilisation of particles sizes during cycling and the creation of a fast diffusion path for hydrogen atoms play a major role (Friedrichs *et al.*, 2006; Friedrichs *et al.*, 2006a).

3.3.1.3.2 Complex hydrides   Other intensely studied materials with high hydrogen content are complex hydrides, for example, $LiBH_4$ (18.4 wt% H_2) (Zuttel *et al.*, 2007; Orimo *et al.*, 2007). In a simplified picture, complex hydrides are salt-like structures of the type $A^{n+}[XH]^{n-}$, where A is either an alkaline, alkaline earth, or early transition metal and X typically is boron or aluminum. The bond between hydrogen and X has a covalent character, and the resulting complex is a rather stable entity which fulfils the '18 electron' rule, representing a closed-shell electronic structure. As for the salt-like binary hydrides, complex hydrides are mostly wide-band-gap insulators with correspondingly poor electrical and thermal conductivity. Complex hydrides were thought to be unsuitable for storage applications due to a lack of reversibility until Bogdanovic and Schwickardi (1997) demonstrated the reversible hydrogen cycling of $NaAlH_4$ doped with a few mole percent of a transition metal salt. Hydrogen release and uptake in $NaAlH_4$ take place in a two-step solid-state reaction:

$$3NaAlH_4 \leftrightarrow Na_3AlH_6 + 2Al + 3H_2 \quad 3.7 \text{ wt\%}, \Delta H = 38 \text{kJ mol } H_2^{-1}$$
$$(3.7a)$$

$$Na_3AlH_6 \leftrightarrow 3NaH + Al + 1.5H_2 \quad 1.7 \text{ wt\%}, \Delta H = 47 \text{kJ/mol } H_2^{-1}$$
$$(3.7b)$$

Additives such as $TiCl_3$, $ScCl_3$, or $CeCl_3$ enable (almost) complete reversible conversion, and rehydrogenation is obtained at 100°C and 10 MPa H_2, while the bare material $NaAlH_4$ exhibits negligible reversibility under these conditions. The release of hydrogen from NaH occurs above 400°C, which is considered impractical for applications. The thermodynamic properties of reactions (3.7a) and (3.7b) are within the region envisioned for solid-state hydrogen storage, and – from a thermodynamic viewpoint – suitable complex hydrides could find application as storage materials.

The effect of the catalyst is still under debate, and while numerous different materials have been identified to be able to catalyse the reaction (Anton, 2003; Fichtner *et al.*, 2003; Graetz *et al.*, 2004; Ichikawa *et al.*, 2005), a unified picture for their activity is still lacking. There is evidence that the transition metal ions are reduced to the metallic state upon cycling (Leon *et al.*, 2007; Leon *et al.*, 2009). Small particle sizes of the additive and homogeneous distribution seem to be beneficial for the performance. Using colloidal $Ti_{13} \cdot$ 6 THF clusters as an additive yields 80% hydrogen uptake in less than 1,000 seconds (at 100°C and 10 MPa H_2). The rate-limiting process is mainly diffusion controlled, which is not surprising in view of the fact that phase segregation of two (or more) solid phases occurs (Lohstroh and Fichtner, 2007).

Besides $NaAlH_4$ other complex hydrides also have been investigated, especially those with higher hydrogen content. So far, reversibility of these compounds is frequently limited or only achieved at harsh conditions. For instance, $LiBH_4$ can be reformed from the elements at 700°C and 150 bar H_2 (Friedrichs *et al.*, 2008). Conditions are slightly milder when LiB_x precursors such as LiB_3 or Li_7B_6 intermetallics are used. Other complex hydrides with limited reversibility include, for example, $Mg(BH_4)_2$ or $Ca(BH_4)_2$, which can be (partially) rehydrogenated at high pressures (90 MPa) and high temperatures (400°C) (Ronnebro and Majzoub, 2007; Severa *et al.*, 2010). Especially, $Mg(BH_4)_2$ is attractive as a storage material since it has a gravimetric storage capacity of 14.9 wt% H_2 and a reported enthalpy of 39 kJ mol H_2^{-1} for the reaction (Matsunaga 2008).

$$Mg(BH_4)_2 \rightarrow MgH_2 + B + 3H_2 \qquad (3.8)$$

So far, the kinetic barriers have not been unambiguously identified which necessitate the high desorption temperatures (>300°C) and hamper rehydrogenation.

Several combinations for (catalytic) additives have been investigated for reversible rehydrogenation of both $Mg(BH_4)_2$ and $Ca(BH_4)_2$ either by reactive ball milling (under hydrogen pressure) or high-temperature and high-pressure treatments. Encouraging results have been obtained for TiF_3 and $TiCl_3/Pd$ mixtures, but more work is needed to understand the role of the additives and structural implications.

3.3.1.3.3 Tailoring thermodynamic properties: reactive hydride composites Beside kinetic limitations, the operating conditions can also be restricted by thermodynamic constraints, for example, the equilibrium decomposition temperature of $LiBH_4$ is 370°C at 0.1MPa H_2 (Mauron *et al.*, 2008). The potential of modified reaction enthalpies is investigated in so called reactive hydride composites (RHCs), that is, composite systems comprised of complex hydrides and a secondary partner. Hydrogen desorption is then accompanied by the formation of a binary compound, thus altering the reaction path and the enthalpy. Originally, the concept was introduced by Reilly and Wishall (1967), who demonstrated reversible hydrogen uptake in the system:

$$3MgH_2 + MgCu_2 \leftrightarrow 2Mg_2Cu + 3H_2 \tag{3.9}$$

The intermetallic compound Mg_2Cu stabilises the desorbed state, and the overall reaction enthalpy decreases but at the expense of a reduced hydrogen storage capacity.

Various reaction based systems combining complex and binary hydrides have been investigated (Vajo *et al.*, 2005; Pinkerton *et al.*, 2007; Barkhordarian *et al.*, 2008), among them, for example, $LiBH_4$ + MgH_2, $Ca(BH_4)_2$ + MgH_2, or $NaBH_4$ + MgH_2. Some examples are given in Table 3.7.

Compared with the pristine hydrides, reversibility in the hydride composites is often improved. Besides thermodynamic alterations, kinetic improvements have also been claimed for borohydride-based RHCs, suggesting facilitated hydrogen uptake from the layered structure of MgB_2 (compared with cage-like elemental boron) (Barkhordarian *et al.*, 2007).

The microstructure of these composites is of crucial importance for the performance and reversibility of the solid-state reaction. For the system $2LiBH_4$ + MgH_2, it was found that Ti-based additions improve the reversibility and enhance reaction rates (Bosenberg *et al.*, 2007). After cycling, nanoscale TiB_2 has been identified as residing between adjacent grains, and it has been suggested that this stable phase acts both as a heterogeneous nucleation centre for the formation of MgB_2 (which is structurally related) and as a grain refiner during cycling, that is, hampering particle growth (Deprez *et al.*, 2010; Bosenberg *et al.*, 2010). Thus dedicated modifications

Table 3.7   Hydrogen storage capacity and reaction enthalpy of some hydrides and RHCs

Reaction	wt% H$_2$	$\Delta H[$ kJ mol^{-1} H$_2]$	Reference
LiH $\leftrightarrow$ Li + 1/2 H$_2$	12.6	181	(Chase, 1998)
NaH $\leftrightarrow$ Na + 1/2 H$_2$	4.2	112.8	(Chase, 1998)
MgH$_2$ $\leftrightarrow$ Mg + H$_2$	7.6	74.5	(Chase, 1998)
3MgH$_2$ + MgCu$_2$ $\leftrightarrow$ 2 Mg$_2$Cu + 3H$_2$	2.6	70.9	(Reilly and Wiswall, 1967)
NaAlH$_4$ $\leftrightarrow$1/3Na$_3$AlH$_6$ + 2/3Al + H$_2$	3.7	38	(Bogdanovic and Schwickardi, 1997)
Na$_3$AlH$_6$ $\leftrightarrow$ 3NaH + Al + 3/2H$_2$	1.5	47	(Bogdanovic and Schwickardi, 1997)
LiBH$_4$ $\leftrightarrow$ LiH + B + 3/2H$_2$	13.9	74	(Mauron *et al.*, 2009)
NaBH$_4$ $\leftrightarrow$ NaH + B + 3/2H$_2$	8.0	135	(Barkhordarian *et al.*, 2008)
2LiBH$_4$+ MgH$_2$ $\leftrightarrow$ MgB$_2$ + 2LiH+4H$_2$	11.5	40.5	(Vajo *et al.*, 2005)
2NaBH$_4$+ MgH$_2$ $\leftrightarrow$ MgB$_2$ + 2NaH + 4H$_2$	7.9	62	(Barkhordarian *et al.*, 2008)
Ca(BH$_4$)$_2$ + MgH$_2$ $\leftrightarrow$ CaH$_2$+ MgB$_2$+ 4H$_2$	8.4	27.5	(Barkhordarian *et al.*, 2008)
LiNH$_2$ + LiH $\leftrightarrow$ Li$_2$NH + H$_2$	6.5	45	(Chen *et al.*, 2002)
Li$_2$NH + LiH $\leftrightarrow$ Li$_3$N + H$_2$	5.4	116	(Chen *et al.*, 2002)
Mg(NH$_2$)$_2$ + 2LiH $\leftrightarrow$ Li$_2$Mg(NH)$_2$	5.6	39	(Luo and Ronnebro, 2005)

of the microstructure could be essential for improving the performance of reaction-based systems. Nevertheless, typical operating temperatures of borohydride systems are still around 400°C.

Nitrogen-based RHCs are especially noteworthy: composites comprised of lithium amide and lithium hydride were reported to store and release hydrogen according to the following reaction (Chen *et al.*, 2002).

$$LiNH_2 + 2LiH \leftrightarrow Li_2NH + LiH + H_2 \qquad (3.9a)$$

$$Li_2NH + LiH + H_2 \leftrightarrow Li_3N + H_2 \qquad (3.9b)$$

In total, 10 wt% hydrogen are released, and the enthalpies of reaction are 65 and 116 kJ mol H$_2^{-1}$ for steps (3.9a) and (3.9b), respectively. Exchanging LiNH$_2$ with the analogue Mg compound Mg(NH$_2$)$_2$ yields a lower stability ($\Delta H = 39$ kJ mol H$_2^{-1}$), and an equilibrium pressure of 0.1 MPa at 90°C was reported for the reaction (Lou and Ronnebro, 2005):

$$Mg(NH_2)_2 + 2LiH \leftrightarrow Li_2Mg(NH)_2 + H_2 \qquad (3.10)$$

The reaction is fully reversible at 200°C and pressures of 10 MPa H_2. Again, the reaction occurs in several steps, which can be clearly distinguished in the pressure composition isotherms (Aoki *et al.*, 2007), and a new phase $Li_2Mg_2(NH)_3$ has been identified to occur alongside $LiNH_2$ and LiH at intermediate temperature and pressure conditions (Weidner *et al.*, 2009). Reaction (3.10) can be written as:

$$2Mg(NH_2)_2 + 4LiH \leftrightarrow Li_2Mg_2(NH)_3 + LiNH_2 + LiH + 3H_2 \quad (3.10a)$$

$$\leftrightarrow 2Li_2Mg(NH)_2 + 4H_2 \qquad (3.10b)$$

The Li-Mg-NH system also constitutes a solid-state reaction but with the peculiarity that the involved phases (with the exception of LiH) are structurally very similar, as can be seen in Figure 3.30.

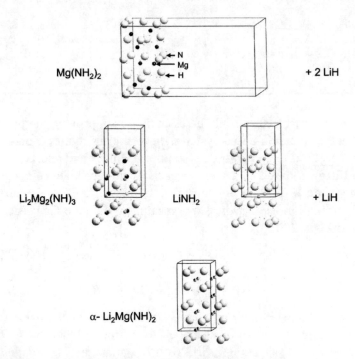

Figure 3.30 Crystal structure of α-$Li_2Mg(NH)_2$, $LiNH_2$, $Li_2Mg_2(NH)_3$, and $Mg(NH_2)_2$. For clarity, only selected parts of the unit cells are shown.

The underlying structure is of anti-fluoride type with nitrogen atoms residing on face-centred cubic (fcc) lattice sites, while the tetrahedral sites are occupied by cations or vacancies. A highly disordered cation lattice has been reported for both the imide and the intermediate phase (Rijssenbeek *et al.*, 2008), and it seems conceivable that the high degree of disorder together with the high density of vacancies facilitates the mass transfer necessary for any solid-state reaction. Doping the systems with potassium compounds such as potassium hydride further lowers the temperatures to 150°C (Wang *et al.*, 2009), and the system thus approaches the operating range for PEM fuel cells.

3.3.1.3.4 Tailoring thermodynamics in small particles Another approach to enhance both reaction rates and possibly adjust thermodynamic properties is the use of particles of limited size. Most prominently, small particles imply a shorter diffusion length and thus higher reaction rates can be expected. In addition, particles with sizes in the lower nanometre range might have different stabilities compared with bulk due to the enhanced surface free-energy contribution, strain, or clamping effects (Berube *et al.*, 2010). For clusters of PdH_x (Pundt, 2004), a shift of phase boundaries has been demonstrated and a diminished two-phase region and lower critical temperature are observed. Similar effects can be observed in thin-film samples (Song *et al.*, 1996; Klose *et al.*, 2000). However, significant deviations from bulk can only be expected for very small particles sizes, for example, a reduction of reaction enthalpy has been predicted for particles sizes below 1.3 nm for the $Mg-MgH_2$ system (Wagemans *et al.*, 2005).

A further challenge is the utilisation of complex hydride particles since upon decomposition these materials segregate into two or more solid phases. That is why isolated particles suffer from insufficient reversibility (Balde *et al.*, 2008, Balde *et al.*, 2006), and the focus shifted to encapsulated systems. Here, the active material is confined into a porous matrix with defined pore dimensions. Encapsulation can be achieved either by melt infiltration (under hydrogen pressure) or by impregnation from a solution. Among the studied materials are $LiBH_4$ and MgH_2 in carbon aerogels (Gross *et al.*, 2009; Gross *et al.*, 2008) and $NaAlH_4$ or $Mg(BH_4)_2$ in carbon matrices (Adelhelm *et al.*, 2010; Fichtner *et al.*, 2009). Experimental results indicate that the equilibrium state can indeed be modified as $NaAlH_4$-carbon composites (with the majority of pore sizes < 4 nm) exhibit altered thermodynamic stability compared with bulk $NaAlH_4$ (Lohstroh *et al.*, 2010).

While the pressure composition isotherm of 'bulk' $NaAlH_4$ (catalysed with either $CeCl_3$ or $TiCl_3$) shows two plateaus corresponding to the two reaction steps, the pressure composition isotherms of $NaAlH_4$ composites point to a broad distribution of thermodynamic properties. In similar systems a reduced stability range for Na_3AlH_6 was reported, while $[AlH_4]^{-1}$ entities could still be evidenced in the absorbed state (Verkuijlen *et al.*, 2010; Adelhelm *et al.*, 2009). Moreover, sodium intercalation into the graphitic layers was reported (Adelhelm *et al.*, 2009). One might therefore speculate that the altered thermodynamic properties are a superposition of small particle effects and a reaction between active material and the host.

3.3.1.4 Materials for energy storage in batteries

Electrochemical energy storage in primary or secondary batteries is a well-established technique used in diverse applications such as in today's consumer electronics, power tools, or starter batteries in cars. It was also used to buffer the electrical grid (e.g., until 1990 in the insular electrical grid in West Berlin, Germany), and currently the use of battery electric cars is intensely discussed.

In general, a battery consists of an anode, an electrolyte, and a cathode (see Figure 3.31). During discharge (positively charged) ions migrate from the anode to the cathode, where they are reduced, while electrons are concurrently transported along the leads.

In secondary or rocking-chair cells, the process can be reversed by an external current and the cell can be cycled many times. The materials' combination determines both the cell voltage and the charge capacity and thus the maximum energy density of the cell. The power density or rate capability is mainly determined by the reaction rate, depending on the velocity of ion migration inside the electrodes and the electrolyte, the number of active reaction sites, and the transfer resistance. The processing and production of the

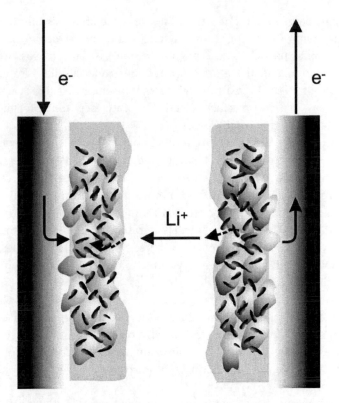

Figure 3.31 Schematic representation of a battery cell. See also Colour Insert.

electrodes are thus crucial for the battery performance, and it has to be optimised for a multitude of factors, for example

- electrical contact of the active material to the current collector,
- ionic current to the electrolyte,
- stability (compatibility) of electrodes in the electrolyte, and
- mechanical stability during cycling as most active materials show considerable volume expansion during ion uptake.

Currently used electrodes usually comprise a mixture of the active materials, a conducting agent such as carbon black, and binders to guaranty homogeneous electrical contact, good soaking of electrolyte, and mechanical stability. Nanostructured electrode designs have a large potential to address the above-listed requirements with short migration distances and many active surface sites. Small particle sizes can also minimise the stresses during intercalation and therefore improve electrode stability. Potential drawbacks of the higher reactivity might be enhanced side reactions such as excessive surface electrolyte interface (SEI) formation counteracting the benefits.

Depending on the type of application, batteries have to comply with different needs. For instance, utilisation in hybrid electrical cars follows a heavy-duty cycle with frequent start/stop cycles requiring high power densities on short time scales followed by brief recharging periods during recuperation of break energy. For some time, this was the domain of nickel–metal hydride (Ni-MH) batteries; however, this cell type is not really suited for full electric vehicles due to their low gravimetric energy density and self-discharge during idle times.

The most promising candidate for electric mobility is lithium ion technology. Li ion secondary batteries are commonly used in today's consumer electronics, usually comprising a carbon-based anode (mainly graphite), a transition metal oxide cathode, and $LiPF_6$ salt dissolved in mixtures of aprotic organic solvents as an electrolyte.

Commonly used cathode materials are $LiCoO_2$, $LiNi_zMn_z$ $Co_{2-2z}O_2$, $LiMn_2O_4$, and $LiFePO_4$ (Whittingham, 2004; Goodenough and Kim, 2010). $LiCoO_2$ has a layered structure, and the practical charge capacity is around 150 mAh g^{-1}. The relative high price for cobalt and its toxicity have triggered the search for alternatives. $LiNi_zMn_zCo_{2-2z}O_2$ is such an successful alternate with lower Co content. In these materials Li resides between adjacent sheets of transition metal ions and oxygen. The conversation of the structural order is essential for the performance, that is, the occupancy of Li sites by transition metal ions has to be minimised.

$LiMn_2O_4$ is a rather cheap and environmentally friendly material, although with a lower capacity of approximately 120 mAh g^{-1}. It suffers from a low cycle life originating from the instability of manganese ions in the electrolytes. Several proposals (Yi *et al.*, 2009) have been made to prevent the dissolution of Mn^{2+} ions: One approach focuses on increasing the structural stability of the lattice by doping the material with various cations such as aluminum or magnesium. Other promising routes aim at minimising the contact area between active materials and the electrolyte by surface coatings or the employment of core shell particles. Potential coating materials include $LiCoO_2$, $LiNi_{0.8}Co_{0.2}O_2$, and nanosize particles of MgO, ZnO, or SiO_2. Encouraging results for improved corrosion resistance have also been obtained with phosphate (e.g., $AlPO_4$) or fluoride coatings. Carbon or metal surface layers (e.g., Ag or Au) have been investigated to enhance the electronic conductivity, while ceramic surface layers were studied for improved ionic conductivity.

$LiFePO_4$ is attractive for its inherent safety, low price, and ecological friendliness. A major challenge for utilisation is the poor electronic and ionic conductivity. Advanced processing techniques enable the preparation of submicron particle sizes with good crystallinity, thus reducing the diffusion path length. Electric conductivity can be enhanced by subsequent coating with carbon (Konstantinov *et al.*, 2004) or electrochemical active polymers (Huang and Goodenough, 2008). Moreover, substitutions with, for example, magnesium, titanium, or zirconium, have been found to improve the rate capabilities (Wang *et al.*, 2004). Commercial Li-$FePO_4$-based electrodes (so far mainly used in power tool applications) still exhibit a rather low packing density of the active material due to the small particle size, and raising the active material density still remains a technological challenge.

The standard material for the negative electrode is graphite with a capacity of 300 mAh g^{-1}. Upon initial formation of the cell, a solid SEI is formed, effectively preventing exfoliation of graphite in subsequent cycles. The low potential of Li intercalation into graphite poses the risk of lithium plating and dendrite formation at the anode.

Alternative anode materials currently under investigation are titanium oxides or the spinel-type $Li_4Ti_5O_{12}$ (Hsiao *et al.*, 2008; Lindsay *et al.*, 2007; Shenouda and Murali, 2008). The latter has negligible volume expansion during lithium insertion, promising less cracking of the electrode and higher cycle life. The cell potential of 1.5 V versus Li prevents Li plating; thus these electrode materials are inherently safe. Problematic are the low ionic and electronic conductivities, requiring small particle sizes. As for other intercalation materials, the challenge is to obtain small particles with good crystallinity.

Titanium oxides exhibit two major polymorphs, anatase and rutile, where the former is usually more electrochemically active. In addition, the morphology can be rather easily controlled by the growth conditions, and a variety of structures such as nanorods and nanowires made from TiO_2 or TiO_2-B have been tested for battery applications (Lan *et al.*, 2004; Armstrong *et al.*, 2004; Armstrong *et al.*, 2005; Armstrong *et al.*, 2006). The small dimensions combined with the high surface area significantly increase the reactivity compared with micron-sized particles.

3.3.1.5 'New' battery materials

In the materials discussed earlier, at most one lithium ion per transition metal atom can be transferred, and therefore only gradual improvements in energy density are to be expected.

However, full electric vehicles compatible with today's mobility characteristics require a larger energy density by a factor of 4 to 5. This will not be possible to realise with current cell chemistries. Similar arguments hold for high-potential cathode materials with open cell voltages >5 V (vs. Li metal) such as $LiCoPO_4$ or $LiMn_{1.5}(Co, Fe, Cr)_{0.5}O_4$. The gain in energy density is directly related to the higher cell voltage, while the number of transferred lithium ions remains more or less constant.

However, new electrolytes with a large electrochemical window have to be developed to assess the full potential of these materials, for example, lithium-conducting ionic liquids (Armand *et al.*, 2009).

For a significant increase of the energy density compared with the currently available systems, new concepts are investigated that enable multi-electron redox reactions. Among the investigated systems are metallic alloys, conversion-based materials, and Li-sulfur or Li-air batteries. The challenges and opportunities of the unconventional cell chemistries will be discussed later in more detail. As for intercalation materials, the efficiency of the electrochemical reaction crucially depends on the number of active sites, the ion mobility, and the electronic contact. Nanotechnology can provide the tools for the materials development of these next-generation battery materials.

3.3.1.5.1 Metal alloys as electrode material   Potential high-capacity anodes based on metal alloys (Thackeray *et al.*, 2003) include tin or silicon (Benedek and Thackeray, 2002; Kasavajjula *et al.*, 2007) with respective charge capacities of 994 mAh g^{-1} and 4,212 mAh g^{-1} corresponding to 4.4 transferred lithium ions per metal atom. The volume expansion during Li uptake presents a significant difficulty imposing capacity fading and a limiting cycle life. The accommodation of mechanical stress during lithium uptake is of outmost importance. One suggestion to minimise cracking and to mitigate mechanical stress is to use small particles in a matrix. For this purpose, mainly composites of nanosized particles (5–100 nm) of silicon (Wang *et al.*, 2004a) or tin (Wang *et al.*, 2004b; Wang *et al.*, 2004c; Derrien *et al.*, 2007; Hassoun *et al.*, 2008) are investigated together with carbon. Various synthesis routes have been used, such as high-energy ball milling, pyrolysis from organic precursors, sol-gel processes, or electrodeposition, all with beneficial effects on capacity retention. In Sn/C composites a reversible capacity of ∼500 mAh g^{-1} has been claimed for several hundred cycles (Hassoun *et al.*, 2008). Other researchers have focused on the preparation of dedicated morphologies and 3D nanostructures of porous silicon. Such structures can be obtained from filling suitable templates (e.g., ordered mesoporous silica) and subsequent leaching of the substrate (Cho, 2010).

Electrodes prepared this way showed 2,800 mAh g^{-1} at a 0.2 C rate, 98% coulombic efficiency, and better capacity retention than nanoparticles of Si or Si-carbon composites. Similarly, Si nanopillars prepared by etching from bulk substrates were studied. This dedicated morphology prevents cracking as it accommodates the volume increase in the available free space, while the pillar geometry presets the expansion direction and maintains good electrical contact to the current collector (Green *et al.*, 2003).

3.3.1.5.2 Conversion-based electrodes   Other concepts to increase electrode capacities focus on conversion materials where the redox reaction is realised in a solid-state reaction of two or more solid

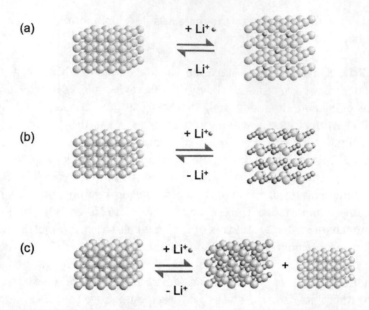

Figure 3.32 Schematic representation of Li uptake in the electrode materials: (a) Li intercalation, (b) alloying, and (c) solid-state conversion reaction. The volume expansion due to Li uptake increases from (a) to (c). See also Colour Insert.

reactants, as schematically depicted in Figure 3.32. Usually more than one electron per transition metal ion can be transferred. Binary transition metal fluorides (such as FeF_3 or TiF_3) (Badway *et al.*, 2003a; Badway *et al.*, 2003b; Li *et al.*, 2003) are investigated as potential cathode materials according to the general reaction (Tm: transition metal):

$$TmF_n + n\,Li \leftrightarrow n\,LiF + Tm \qquad (3.11)$$

Due to the highly ionic bonds of fluoride salts, the open cell potentials are in the range of 2–4 V versus Li, and all three reduction/oxidation steps can be realised in the electrode reaction (e.g., for FeF_3 or TiF_3). The theoretical capacities are $> 700\ \mathrm{mAh\ g^{-1}}$ (see Table 3.8); thus these materials are attractive as high-capacity positive electrodes.

Early experiments suffered from a low cycle life and slow kinetics. The problems at hand are the high volume change during the reaction, leading to disintegration of the electrode; in addition,

Table 3.8   Theoretical cell voltage and specific capacity for selected metal fluorides (Li *et al.*, 2003)

	E[V] vs Li	C [mAhg^{-1}]
FeF_3	2.74	712
CoF_3	3.62	694
CuF_2	3.55	528
NiF_2	2.96	554
VF_3	1.86	745
ZnF_2	2.40	518

electronic conductivity of fluorides is extremely poor. The material development therefore has to provide for a small diffusion path length, good electrical contact, high (ion) mobility of all the solid reactants, and mechanical stability.

Small particles sizes of active materials embedded in a conducting matrix have therefore been in the centre of attention. Composites of transition metal fluorides with carbon prepared by high-energy ball milling can be reversibly cycled.

Encouraging results have been also obtained for nanosized CuF_2 (2–30 nm) particles embedded into a metal oxide matrix with mixed ionic and electronic conductivity (Badway *et al.*, 2007). Using a new synthesis approach for the charged state, nanosized Fe particles of around 10 nm have been encapsulated in carbon fibre structures mixed with finely dispersed LiF particles (Prakash *et al.*, 2010). Such composites have been demonstrated to have a capacity of 280 mAh g^{-1} (in the voltage range 4.3–0.5 V), small capacity fading, and good rate capabilities.

Examples for conversion reactions are widespread, including also oxides, sulfides, nitrides, or phosphates (Poizot *et al.*, 2000; Debart *et al.*, 2006; Gillot *et al.*, 2005; Laruelle *et al.*, 2002). These compounds are rather investigated as anode materials, owing to the lower open circuit potential (vs. Li). Depending on the oxidation state of the transition metal, one or more electrons are transferred, for example, two in CoO, three in Fe_2O_3, and four in RuO_2. Compared with the currently used graphite, their cell voltage is slightly higher, which reduces the risk of Li plating on the negative electrode and thus increases the safety of the cell (Bruce *et al.*, 2008).

Systems promising much higher energy densities compared with the currently available techniques are Li-air or Li-sulfur cells (with theoretical capacities of 1,170 mAh g^{-1} and 1675 mAh g^{-1}, for Li-air and Li-sulfur, respectively [Gao and Yang, 2010]), but so far, there remain a number of difficulties to be solved for large-scale applications of these cell chemistries. One of the main problems attached to Li-sulfur cells is the incompatibility with conventional electrolytes as the reaction products LiS_x are solvable in the currently used organic liquids. Correspondingly reversibility is limited. A solution could be the employment of gel-polymer electrolytes or ionic liquids with high enough Li mobility but minimised solubility.

Li-air cells suffer from an opposite problem, that is, Li ions travel through the electrolyte to the cathode side, where the reaction with gaseous oxygen yields a solid reaction product, LiO_x potentially blocking the pores of the separator membrane. Moreover, the round-trip efficiency still has to be improved, and effective catalysts are needed for O_2 splitting and recombination.

3.3.1.6 Conclusions

The development of materials for energy storage has made great advances in recent years. One of the important factors for the improved performance of fuel cells, solid-state hydrogen storage, or batteries is the ability to control particle sizes, composition, and morphology on the nanoscale.

Efficient energy storage is required in many, quite diverse applications with no unique solution equally well suited for all requirements. The utilised materials in the various applications cover a wide range, and different processing techniques are required. Nevertheless, a common feature to raise the capabilities of any electrochemical energy storage system will be the ability to tailor the reaction sites in solid-solid, solid-liquid, or solid-gas reactions. For high rates, the efficient transport of the reactants towards the reaction site is important, that is, electron and ion mobility in case of batteries, hydrogen transport in solid-state hydrogen storage materials, but also materials transport in solid-state reaction systems such as conversion materials. On the other hand, dedicated designs can improve mechanical stability of the macroscopic electrodes or serve as a protective coating to enhance the compatibility of the various components. In all these areas, nanotechnology will significantly contribute to next-generation energy storage materials.

Bibliography

Adelhelm, P., de Jong, K. P., and de Jongh, P. E. (2009). *Chem. Commun.*, 6261.

Adelhelm, P., Gao, J., Verkuijlen, M. H., Rongeat, C., Herrich, M., van Bentum, J. M., Gutfleisch, O., Kentgens, A. O. M., de Jong, K., and de Jongh, P. (2010). *Chem. Mater.*, 22, 2233.

Aguey-Zinsou, K. F. and Ares-Fernandez, J. R. (2010). *Energy Environ. Sci.*, 3, 526, and references therein.

Anton, D. L. (2003). *J. Alloys Compd.*, 356, 400.

Aoki, M., Noritake, T., Nakamori, Y., Towata, S., and Orimo, S. (2007). *J. Alloys Compd.*, 446, 328.

Armand, M., Endres, F., MacFarlane, D. R., Ohno, H., and Scrosati, B. (2009). *Nat. Mater.*, 8, 621.

Armstrong, A. R., Armstrong, G., Canales, J., and Bruce, P. G. (2004). *Angew. Chem., Int. Ed.*, 43, 2286.

Armstrong, A. R., Armstrong, G., Canales, J., Garcia, R., and Bruce, P. G. (2005). *Adv. Mater.*, 17, 862.

Armstrong, G., Armstrong, A. R., Canales, J., and Bruce, P. G. (2006). Electrochem. *Solid State Lett.*, 9, A139.

Badway, F., Mansour, A. N., Pereira, N., Al-Sharab, J. F., Cosandey, F., Plitz, I., and Amatucci, G. G. (2007). *Chem. Mater.*, 19, 4129.

Badway, F., Pereira, N., Cosandey, F., and Amatucci, G. G. (2003a). *J. Electrochem. Soc.*, 150, A1209.

Badway, F., Pereira, N., Cosandey, F., and Amatucci, G. G. (2003b). Solid State Ionics-2002, 756, 207.

Balde, C. P., Hereijgers, B. P. C., Bitter, J. H., and de Jong, K. P. (2006). *Angew. Chem., Int. Ed.*, 45, 3501.

Balde, C. P., Hereijgers, B. P. C., Bitter, J. H., and de Jong, K. P. (2008). *J. Am. Chem. Soc.*, 130, 6761.

Barkhordarian, G. Jensen, T. R., Doppiu, S. Bösenberg, U., Borgschulte, A., Gremaud, R., Cerenius, Y., Dornheim, M., Klassen, T., and Bormann, R. (2008). *J. Phys. Chem.* C, 112, 2743.

Barkhordarian, G., Klassen, T., and Bormann, R. (2006). *J. Phys. Chem.* B, 110, 11020.

Barkhordarian, G., Klassen, T., Dornheim, M., and Bormann, R. (2007). *J. Alloys Compd.*, 440, L18.

Benedek, R. and Thackeray, M. M. (2002). *J. Power Sources*, 110, 406.

Berube, V., Radtke, G., Dresselhaus, M., and Chen, G. (2007). *Int. J. Energy Res.*, 31, 637.

Bogdanovic, B. and Schwickardi, M. (1997). *J. Alloys Compd.*, 253, 1.

Bosenberg, U, Kim, J. W., Gosslar, D., Eigen, N., Jensen, T. R., Bellosta von Colbe, J. M. Zhou, Y., Dahms, M., Kim, D. H., Günther, R., Cho, Y. W., Oh, K. H., Klassen, T., Bormann, R., and Dornheim, M. (2010). *Acta Mater.*, 58, 3381.

Bosenberg, U., Doppiu, S., Mosegaard, L., Barkhordarian, G., Eigen, N., Borgschulte, A., Jensen, T. R., Cerenius, Y., Gutfleisch, O., Klassen, T., Dornheim, M., and Bormann, R. (2007). *Acta Mater.*, 55, 3951.

Bruce, P. G., Scrosati, B., and Tarascon, J. M. (2008). *Angew. Chem., Int. Ed.*, 47, 2930.

Chase, M. W. (1998). NIST-JANAF themochemical tables, Fourth Edition, *J. Phys. Chem. Ref. Data, Monograph* 9, 1-1951.

Chen, P., Xiong, Z. T., Luo, J. Z., Lin, J. Y., and Tan, K. L. (2002). *Nature*, 420, 302.

Cho, J. (2010). *J. Mater. Chem.*, 20, 4009.

Debart, A., Dupont, L., Patrice, R., and Tarascon, J. M. (2006). *Solid State Sci.*, 8, 640.

Deprez, E., Munoz Márques, M. A., Roldán, M. A., Prestipino, C., Plaömares, F. J., Bonatto, C., Bösenberg, U., Dornheim, M., Bormann, R., and Fernández, A. (2010). *J. Phys. Chem.* C, 114, 3309.

Derrien, G., Hassoun, J., Panero, S., and Scrosati, B. (2007). *Adv. Mater.*, 19, 2336.

Dornheim, M., Doppiu, S., Barkhordarian, G., Boesenberg, U., Klassen, T., Gutfleisch, O., and Bormann, R. (2007). *Scr. Mater.*, 56, 841.

Dornheim, M., Eigen, N., Barkhordarian, G., Klassen, T., and Bormann, R. (2006). *Adv. Eng. Mater.*, 8, 377.

Eberle, U., Felderhoff, M., and Schuth, F. (2009). *Angew. Chem., Int. Ed.*, 48, 6608.

Fichtner, M., Fuhr, O., Kircher, O., and Rothe, J. (2003). *Nanotechnology*, 14, 778.

Fichtner, M., Zhao-Karger, Z., Hu, J. J., Roth, A., and Weidler, P. (2009). *Nanotechnology*, 20(2009).

Friedrichs, O., Buchter, F., Borgschulte, A., Remhof, A., Zwicky, C. N., Mauron, P., Bielmann, M., and Zuttel, A. (2008). *Acta Mater.*, 56, 949.

Friedrichs, O., Klassen, T., Sanchez-Lopez, J. C., Bormann, R., and Fernandez, A. (2006). *Scr. Mater.*, 54, 1293.

Friedrichs, O., Sánchez-López, J. C., López-Cartes, C., Klassen, T., Bormann, R., and Fernández, A. (2006). *J. Phys. Chem.* C, 110, 7845.

Gao, X. P. and Yang, H. X. (2010). *Energy Environ. Sci.*, 3, 174.

Gillot, F., Boyanov, S., Dupont, L., Doublet, M. L., Morcrette, A., Monconduit, L., and Tarascon, J. M. (2005). *Chem. Mater.*, 17, 6327.

Goodenough, J. B. and Kim, Y. (2010). *Chem. Mater.*, 22, 587, and references therein.

Graetz, J., Reilly, J. J., Johnson, J., Ignatov, A. Y., and Tyson, T. A. (2004). *Appl. Phys. Lett.*, 85, 500.

Green, M., Fielder, E., Scrosati, B., Wachtler, M., and Moreno, J. S. (2003). Electrochem. *Solid State Lett.*, 6, A75.

Gross, J. A. F., Ahn, C. C., Van Atta, S. L., Liu, P., and Vajo, J. J. (2009). *Nanotechnology*, 20.

Gross, J. A. F., Vajo, J. J., Van Atta, S. L., and Olson, G. L. (2008). *J. Phys. Chem.* C, 112, 5651.

Gross, K. J., Chartouni, D., Leroy, E. Zuttel, A., and Schlapbach, L. (1998). *J. Alloys Compd.*, 269, 259.

Gross, K. J., Spatz, P., Zuttel, A., and Schlapbach, L. (1997). *J. Alloys Compd.*, 261, 276.

Hassoun, J., Derrien, G., Panero, S., and Scrosati, B. (2008). *Adv. Mater.*, 20, 3169.

Hirscher, M., Panella, B., and Schmitz, B. (2010). *Microporous Mesoporous Mater.*, 129, 335.

Hsiao, K. C., Liao, S. C., and Chen, J. M. (2008). *Electrochim. Acta*, 53, 7242.

Huang, Y. H. and Goodenough, J. B. (2008). *Chem. Mater.*, 20, 7237.

Huot, J., Liang, G., and Schulz, R. (2001). *Appl. Phys. a-Mater. Sci. Proc.*, 72, 187.

Ichikawa, T., Hanada, N., Isobe, S., and Leng, H. Y. (2005). *Mater. Trans.*, 46, 1.

Kasavajjula, U., Wang, C. S., and Appleby, A. J. (2007). *J. Power Sources*, 163, 1003.

Klose, F., Rehm, C., Fieber-Erdmann, M., Holub-Krappe, E., Bleif, H. J., Sowers, H., Goyette, R., Troger, L., and Maletta, H. (2000). *Physica B*, 283, 184.

Konstantinov, K., Bewlay, S., Wang, G. X., Lindsay, M., Wang, J. Z., Liu, H. K., Dou, S. X., and Ahn, J. H. (2004). *Electrochim. Acta*, 50, 421.

Lan, Y., Gao, X. P., Zhu, H. Y., Zheng, Z. F., Yan, T. Y., Wu, F., Ringer, S. P., and Song, D. Y. (2005). *Adv. Funct. Mater.*, 15, 1310.

Langmi, H. W., Walton, A., Al-Mamouri, M. M., Johnson, S. R., Book, D., Speight, J. D., Edwards, P. P., Gameson, J., Anderson, P. A., and Harris, I. R. (2003). *J. Alloys Compd.*, 356, 710.

Laruelle, S., Grugeon, S., Poizot, P., Dolle, M., Dupont, L., and Tarascon, J. M. (2002). *J. Electrochem. Soc.*, 149, A627.

Leon, A., Rothe, J., and Fichtner, M. (2007). *J. Phys. Chem.* C 111, 16664.

Leon, A., Rothe, J., Chlopek, K., Zabara, O., and Fichtner, M. (2009). *Phys. Chem. Chem. Phys.*, 11, 8829.

Li, H., Richter, G., and Maier, J. (2003). *Adv. Mater.*, 15, 736.

Lindsay, M. J., Blackford, M. G., Attard, D. J., Luca, V., Skyllas-Kazacos, M., and Griffith, C. S. (2007). *Electrochim. Acta*, 52, 6401.

Lohstroh, W. and Fichtner, M. (2007). *Phys. Rev.* B, 75(2007), 184106.

Lohstroh, W., Roth, A., Hahn, H., and Fichtner, M. (2010). Chemphyschem, 11, 789.

Luo, W. and Ronnebro, E. (2005). *J. Alloys Compd.*, 404, 392.

Matsunaga, T., Buchter, F., Mauron, P., Bielman, M., Nakamori, Y., Orimo, S., Ohba, N., Miwa, K., Towata, S., and Zuttel, A. (2008). *J. Alloys Compd.*, 459, 583.

Mauron, P., Buchter, F., Friedrichs, O., Remhof, A., Bielmann, M., Zwicky, C. N., and Zuttel, A. (2008). *J. Phys. Chem.* B 112, 906.

Mori, D. (2005). IPHE International Hydrogen Storage Technology Conference, Lucca, Italy, June 19-22, 2005.

Orimo, S. I., Nakamori, Y., Eliseo, J. R., Zuttel, A., and Jensen, C. M. (2007). *Chem. Rev.*, 107, 4111.

Panella, B., Hirscher, M., Putter, H., and Muller, U. (2006). *Adv. Funct. Mater.*, 16, 520.

Pinkerton, F. E., Meyer, M. S., Meisner, G. P., Balogh, M. P., and Vajo, J. J. (2007). *J. Phys. Chem.* C 111, 12881.

Poizot, P., Laruelle, S., Grugeon, S., Dupont, L., and Tarascon, J. M. (2000). *Nature*, 407, 496.

Prakash, R., Mishra, A. K., Roth, A., Kubel, C., Scherer, T., Ghafari, M., Hahn, H., and Fichtner, M. (2010). *J. Mater. Chem.*, 20, 1871.

Pundt, A. (2004). *Adv. Eng. Mater.*, 6, 11, and references therein.

Reilly, J. J., and Wiswall, R. H. (1967). *Inorg. Chem.*, 6, 2220.

Rijssenbeek, J., Gao, Y., Hanson, J., Huang, Q., Jones, C., and Toby, B. (2008). *J. Alloys Compd.*, 454, 233.

Ronnebro, E. and Majzoub, E. H. (2007). *J. Phys. Chem.* B 111, 12045.

Rowsell, J. L. C., Millward, A. R., Park, K. S., and Yaghi, O. M. (2004). *J. Am. Chem. Soc.*, 126, 5666.

Severa, G., Ronnebro, E., and Jensen, C. M. (2010). *Chem. Commun.*, 46, 421.

Shenouda, A. Y., and Murali, K. (2008). *J. Power Sources*, 176, 332.

Song, G., Geitz, M., Abromeit, A., and Zabel, H. (1996). *Phys. Rev.* B, 54, 14093.

Thackeray, M. M., Vaughey, J. T., Johnson, C. S., Kropf, A. J., Benedek, R., Fransson, L. M. L., and Edstrom, K. (2003). *J. Power Sources*, 113, 124.

Thomas, K. M. (2007). *Catal. Today*, 120, 389.

Thomas, K. M. (2009). *Dalton Trans.*, 1487(2009).

Vajo, J. J., Skeith, S. L., and Mertens, F. (2005). *J. Phys. Chem.* B 109, 3719.

Verkuijlen, M. H. W., Gao, J. B., Adelhelm, P., van Bentum, P. J. N., de Jongh, P. E., and Kentgens, A. P. M. (2010). *J. Phys. Chem.* C, 114, 4683.

Wagemans, R. W. P., van Lenthe, J. H., de Jongh, P. E., van Dillen, A. J., and de Jong, K. P. (2005). *J. Am. Chem. Soc.*, 127, 16675.

Wang, G. X., Ahn, J. H., Yao, J., Bewlay, S., and Liu, H. K. (2004a). *Electrochem. Commun.*, 6, 689.

Wang, G. X., Bewlay, S., Yao, J., Ahn, J. H., Dou, S. X., and Liu, H. K. (2004). *Electrochem. Solid State Lett.*, 7, A503.

Wang, G. X., Yao, J., Ahn, J. H., Liu, H. K., and Dou, S. X. (2004b). *J. Appl. Electrochem.*, 34, 187.

Wang, G. X., Yao, J., Liu, H. K., Dou, S. X., and Ahn, J. H. (2004c). *Electrochim. Acta*, 50, 517.

Wang, J., Liu, T., Li, W., Liu, Y., Araujo, M., Scheicher, R. H., Blomqvist, A., Ahuja, R., Xiong, Z. Yang, P., Gao, M., Pan, H., and Chen, P. (2009). *Angew. Chem., Int. Ed.*, 48, 5828.

Weidner, E., Dolci, F., Hu, J. J., Lohstroh, W., Hansen, T., Bull, D. J., and Fichtner, M. (2009). *J. Phys. Chem.* C 113, 15772.

Whittingham, M. S. (2004). Chem. Rev., 104, 4271, and references therein.

Wong-Foy, A. G., Matzger, A. J., and Yaghi, O. M. (2006). *J. Am. Chem. Soc.*, 128, 3494, 2006.

Yi, T. F., Zhu, Y. R., Zhu, X. D., Shu, J., Yue, C. B., and Zhou, A. N. (2009). *Ionics*, 15, 779, and references therein.

Zuttel, A., Borgschulte, A., and Orimo, S. I. (2007). *Scr. Mater.*, 56, 823.

Zuttel, A., Sudan, P., Mauron, P., Kiyobayashi, T., Emmenegger, C., and Schlapbach, L. (2002). *Int. J. Hydrogen Energy*, 27, 203.

3.4 Energy Use

3.4.1 *Nanotechnology in Construction*

Dr. Wenzhong Zhu
Scottish Centre for Nanotechnology in Construction Materials, School of Engineering and Science, University of the West of Scotland, UK

Construction is among the most important industrial sectors in the world, accounting for about 10–11% of the gross domestic product (GDP), 7% of the total employment, and 50–60% of the gross fixed capital formation. It also has significant impact on our lives and societies through housing, infrastructure, use of materials/ energy, and sustainability. However, advances in productivity of the construction industry in the twentieth century were slow, and development of its technology (materials and processes) lagged behind other industrial sectors. As a powerful, emerging enabling tool, nanotechnology is in the position to assist achieving real competitive and sustainable growth and innovation within the construction industry. Applications of nanotechnology are expected to lead to better, smarter, and more durable materials and more sustainable products/processes.

3.4.1.1 General development

The construction industry was one of the first industrial sectors to identify nanotechnology as a promising emerging technology in the 1990s (Gann, 2002; Flanagan *et al.*, 2010; DTI, 2001). Two sub-sectors, that is, ready-mixed concrete and concrete products, were identified in the early 2000s as among the top 40 industrial sectors likely to be influenced by nanotechnology in 10–15 years (In Realis, 2002).

As a powerful, emerging enabling tool, nanotechnology offers the potential to drastically improve the performance (e.g., strength, durability, energy consumption, recyclability, and other special functional properties) of construction materials and to create new materials, sensors/devices, and more sustainable products/production processes for the construction industry.

From 2001, the European Union (EU), the United States, and many other national research funding authorities added construction into their nanotechnology-related portfolios of strategic support. In 2002, the European Commission funded a growth project GMA1-2002-72160 'NANOCONEX' – Towards the Setting Up of a Network of Excellence in Nanotechnology in Construction (Porro, 2004) to support transfer of knowledge in nanotechnology into construction research and technological development (RTD). As part of this project, the first state-of-the-art report on the application of nanotechnology in construction (Zhu, 2004) and a road map for future development of nanotechnology in construction (Bartos, 2009) were prepared.

Nanotechnology research and application in the broad area of construction and the built environment started from a very low base point, with the level of investment in RTD and the awareness of nanotech innovation being the lowest across the whole of the industrial sectors. Surveys carried out in the early 2000s (Gann, 2002; Zhu, 2004) showed that nanotechnology application in construction was a very small, fragmented pursuit and unknown outside the scientific circle. There were generally a lack of awareness and negative perceptions of nanotechnology among construction professionals. Although there were some research activities on new materials and sensors/devices in the construction sector, the linking of a highly complex, breakthrough technology and a sector generally perceived as 'low tech' was considered contradictory (Porro, 2004).

Over the past few years, in line with a very rapid growth in awareness and interest in nanotechnology development in other industrial sectors, the situation in the construction industry and the built environment, particularly in the functional coating and the construction materials sub-sectors, has been improving steadily.

For example, there has been a significant increase in the number of funded nanotechnology-related research projects in the EU, the

United States, and many other countries in the field of construction/construction materials. Several further reports and review papers on the development and application of nanotechnology in construction have been published (Mann, 2006; Zhu, 2007; Freedonia, 2007; Andersen, 2007; Luther, 2009; Sanchez, 2010). International symposiums and workshops on the topics have also become more frequent and widespread. Listed below are some of the major events/meetings which have demonstrated the increasing interests/activities in nanotechnology applications in construction.

- RILEM TC 197-NCM: Nanotechnology in construction materials, established in 2002
- RILEM TC-NBM: Nanotechnology-based bituminous materials, established in 2008
- International symposium on nanotechnology in construction, first in Paisley, 2003; second in Bilbao, 2005; third in Prague, 2009
- Federal highway administration (FHWA)/NSF nanotechnology workshop, USA, 2003
- Euromat on nanoscience of cement materials, Switzerland, 2003
- Nanotechnology for construction materials workshop, Canada, 2004
- Workshop on cementitious materials as model porous media: Nanostructure and transport processes, Switzerland, 2005
- NSF Workshop on Nanotechnology for Cement and Concrete, first in Florida, 2006; second in Washington, DC, 2007
- Workshop on nanotechnology for environmentally friendly construction, Copenhagen, Nanotech Northern Europe, 2008
- 1st International Conference in North America on Nanotechnology in Cement and Concrete, California, May 2010
- Nano Cement, Steel and Construction Industries Conference, Cairo, Egypt, first in May 2009; second in May 2011
- Many other workshops/conferences under MRS, E-core, EMRS, ACI, and specific research projects, etc.

Commercial applications of nanotechnology-based materials/products in construction, however, are still rather limited compared with many other industrial sectors.

Early expectations of a more practical exploitation have not been fulfilled; much more remains to be done (Bartos, 2009). It was reported (Freedonia, 2007) that use of nanomaterials in construction in the United States totaled less than $20 million in 2006, and the market is projected to reach $100 million in 2011 and approach $1.75 billion in 2025. The main research activities and advances to date have been in the generation of new knowledge rather than practical application (Bartos, 2009).

Some of the main obstacles to commercial application of nano-based materials/products in the construction industry have included high initial cost of nano-based materials/products, inherent nature of conservatism, tight regulation, long-term performance requirement, and very low level of industrial involvement in RTD, etc, in the construction industry. On the other hand, the increasing emphasis/requirements on reducing energy consumption and greenhouse gas emission, sustainable development, consideration of the whole life cycle cost, and cross-border competition have become important drivers for nanotechnology application in construction and the built environment. In some cases (e.g., drastically improving the insulation performance of existing buildings/houses), nanotechnology-based solutions are perceived as among the most feasible.

3.4.1.2 Application areas

Nanotechnology is intrinsically multidisciplinary, and its application encompasses many areas of science and engineering. Application of nanotechnology often offers breakthroughs, but at present, the main benefit of using nanotechnology is adding significant extra value to existing solutions in most cases, for example, construction materials/products. Nanotechnology-related RTD and advances in the construction industry have been uneven. Although a limited number of nano-based materials/products has reached the market, the main advances have been in the generation of new knowledge, for example, fundamental understanding of construction materials and control of their nano-/microstructures.

3.4.1.2.1 Bulk materials Bulk construction materials, including cement and concrete, wood, steel, asphalt, ceramics, glass, polymers, and composites, are used on a large scale and produced in huge quantities. To give just one example, at present more than 2 tons of concrete per person are used annually around the world (being the second-most consumed material, only behind water). Application of nanotechnology-enabled new characterisation and modelling tools in studying such traditional materials at increasingly smaller scales could open up the possibility of real breakthroughs in our understanding and control of different mechanisms/processes at the nanoscale. Furthermore, recent work on using nanostructured materials (CNTs, nano-additives, thin films, etc.) to modify microstructures and to develop materials with specific characteristics has shown promising results. Currently, though nano-related RTD activities exist in almost all the above materials sectors, the development and application are very uneven.

Due to their undisputed importance in construction, cementitious materials or concrete have attracted considerable attention and research activities. Cementitious materials are complex, heterogeneous composite materials, with a random microstructure at different length scales, from the nano- to the macro scale. Such complex composites are made even more complicated by the time-dependent nature of the cement hydration processes which begin at the mixing of cement clinker minerals with water and continue for months and even years. The ultimate performance of such highly heterogeneous composite materials is governed by many different (i.e., mechanical, hygral, thermal, and chemical) processes, which are still to be fully understood at the different length scales. Existing nanotechnology-related RTD activities in cementitious materials can be generally divided into the following three areas:

- Fundamental understanding through advanced characterisation and modelling: Advanced instruments, characterisation, and modelling techniques sensitive to micro-/nanoscale structures and properties are increasingly used to study the various structures, properties, and underlying mechanisms of cement-based materials, such as nano-/microstructures and properties of calcium silicate hydrate (CSH) phases, the evolution of the pore structure, and its interactions with moisture/fluids during the course of hydration and degradation, as well as structure-properties relationship, etc.

- Development of enhanced properties/new materials through modification of nano-/microstructures: Research programmes have emerged around the world in the last few years to develop materials with much enhanced performance and functionality (e.g., strength, toughness, durability, insulation, and self-healing) or a set of specific properties required for use in extreme service conditions. Concrete can be successfully modified from the nano-/microscale through the use of suitable nanoparticles, CNTs, nanofibres, and nano-based admixtures/additives (Sanchez, 2010). For example, nano-based polycarboxylate ethers superplasticisers and amorphous nanosilica dispersions (both have been commercially applied) can be used to produce self-compacting concrete, ultra-high-strength concrete, and special concrete with drastically enhanced durability performance. CNTs added to cement matrix could offer higher performance by impeding crack formation and growth. Even a tiny amount of CNT (0.025–0.08 wt% of cement) can significantly increase the strength and stiffness as well as reduce the autogenous shrinkage of the cementitious matrix (Konsta-Gdoutos, 2010). Using suitable nanoscale viscosity modifiers in concrete has also shown to significantly reduce the ion mobility, thus potentially doubling the service life of concrete structures (Bentz, 2008).

- Development of sustainable concrete materials: The manufacturing of cement is an energy-intensive process and is responsible for approximately 5% of the global carbon dioxide emissions. Significant RTD efforts have been made to develop special low-energy and low-emission cements/alternative binders, to use supplementary cementitious materials, and to recycle wastes in concrete, as well as to develop high-performance materials with extended durability or service life (including self-healing characteristics). Nano-modification and use of nano-additives or admixtures offer an important tool for developing green and sustainable concrete materials and structures.

Similar RTD activities also exist in other construction materials, such as steel, ceramics, plastics, composites, and bituminous materials. One of the most significant examples of commercial development in the area is perhaps the nanostructure-modified steel reinforcement – MMFX2 Steel, manufactured by MMFX Steel Corp., USA (MMFX, 2010). The steel has a corrosion resistance similar to that of stainless steel but at a much lower cost. Compared with conventional carbon steel, MMFX steel has a completely different structure at nanoscale – a laminated lath structure resembling 'plywood'. Due to the modified nanostructure, MMFX steel also has superior mechanical properties, for example, much higher strength, ductility, and fatigue resistance. Since corrosion of steel is one of the most serious and costly problems facing construction today, the implications of highly-corrosion-resistant steel, such as MMFX, are far reaching. A large number of projects using MMFX steel has been completed in the United States and Canada.

3.4.1.2.2 Functional coatings/surfaces This area is the most commercially developed/researched, and the use of these nanoscale materials has been the basis for many current and potential applications in construction and the built environment. According to a recent Freedonia report (Freedonia, 2007), coatings are expected to constitute the largest application for nanomaterials in construction. Incorporating certain nanoparticles or nanolayers into coatings/paints and thin films can add enhanced performance and additional functionalities to product surfaces, which can include transparency, photo-reactivity, anti-bacterial property, self-cleaning, scratch resistance, anti-corrosion property, UV blocking, colour durability, and improved thermal performance.

Currently, a considerable number of products have been developed and are commercially available for construction and the built environment. For example, self-cleaning coating/paint based on the photocatalytic properties of titanium dioxide (TiO_2) nanoparticles has been applied to a range of products from self-cleaning glass and cement to interior/exterior wall paints, roof tiles, facades, and road pavements (Pilkington, 2010; Bioni, 2010; Italcementi 2010).

Upon UV (or visible light in some cases) radiation, the nanoparticles TiO_2-based active surface can break down dirt and air pollutants through the production of free radicals. As a result, in addition to self-cleaning, such coatings/paints can also improve the air quality indoors/outdoors. Nanoparticles of SiO_2, Al_2O_3, ZnO, and Ag have also been used to produce special functions (e.g., enhanced durability, scratch resistance anti-microbial property, and anti-bacterial property) in the paints/coatings developed (Bioni, 2010).

Self-cleaning can also be achieved by creating a superhy-drophobic surface–mimicking the self-cleaning function of lotus leaves. BASF has developed a lotus spray for wood surfaces (BASF, 2010), which contains a mixture of silica or alumina nanoparticles, hydrophobic polymers, and a propellant gas, offering 20 times more water-repellent property than a smooth wax coating. Similar products, such as lotus-effect facade paints (Lotus, 2010) and anti-graffiti coating (Castano, 2010) for buildings/structures have also been developed.

There are also a number of promising research activities in insulating, solar cells, smart and responsive coatings for energy-efficient buildings, sensors, and intelligent structures. For example, the development of smart windows through electrochromic, thermochromic, and photochromic coatings, which are capable of intelligently controlling light and heat flows into the building, offers significant potentials in saving energy for heating/cooling and lighting (ChromoGenics, 2010).

3.4.1.2.3 Insulation and energy   Buildings are responsible for 40% of energy consumption and 36% of CO_2 emissions in the EU (EU, 2010). Improving the energy performance of buildings is a cost-effective way of fighting against climate change and improving energy security. In residential buildings most of the energy used is required for domestic hot water and space heating, ventilation, lighting, and cooling. Heat loss and gain are closely connected to the presence of glass surfaces and to the insulating capacity of the outer cladding (Scalisi, 2009). Ambitious targets to cut energy consumption and emission have been set by the EU and other national governments, and the following three strategic steps may be taken to achieve the long-term goal (EC, 2010).

- Step 1: Reducing the energy use of buildings and its negative impacts on environment,

- Step 2: Buildings covering their own energy needs, and
- Step 3: Transforming buildings into energy providers, preferably at the district level.

Nanotechnology is seen to offer huge potentials to meet the challenges through innovative insulating materials, smart windows, innovative lighting, and energy systems. For example, adding thermal insulation to existing European buildings could cut current building energy costs and carbon emissions by 42% or 350 million metric tons (Elvin, 2007). A special programme of research funding has recently been made available on energy-efficient buildings, including new nanotechnology-based high-performance insulation systems for energy efficiency (EC, 2010).

Nanotechnology can help to develop novel insulation materials of amazing properties. Aerogel is such an example, which is an extremely low-density solid with 95% nanosized pores. Aerogels are good thermal insulators because they almost nullify the three methods of heat transfer (convection, conduction, and radiation). As a result, they can be used to produce nearly translucent insulating panels, and a 9 mm thick aerogel achieves the thermal performance of 50 mm of mineral wool (Scalisi, 2009).

Nanotechnology-based insulation coating/paint represents another category of promising materials for building insulation. For example, Nansulate® has developed a series of coating/paint products based on a nanocomposite called hydro-NM-oxide (Nansulate, 2010). The multifunctional clear coatings, which can be applied like paint to different surfaces (interior/exterior walls, windows, pipes, roofs, etc.), provide an effective and affordable means to increase energy efficiency in pre-existing buildings as well as new construction. Test results show that Nansulate® applied in residential buildings achieves a 46% energy saving on average (Nansulate, 2010). One obvious and important advantage of such coatings/paints is their easy application to existing buildings, where a major problem for building renovation is a lack of space for conventional thermal insulation.

Vacuum insulation panels (VIPs), initially developed for refrigerators, have also attracted much attention for building applications. Their insulation performance is a factor of five to ten times better than that of conventional insulation. Used in buildings they enable thin, highly insulating constructions to be realised for walls, roofs, and floors (Nano, 2010).

Nanotechnology also provides potential for more efficient processes of energy supply and lighting in building engineering. Nanostructured catalysts and membranes can offer enhanced efficiency in fuel cell technology. In the field of PV, dye-sensitised solar cells based on dye-doped titanium dioxide nanoparticles provide new design possibilities in building engineering. In the field of lighting engineering, the development and application of energy-efficient light-emitting diodes (LEDs) based on inorganic/organic semiconductor materials have also shown promises (Luther, 2009).

3.4.1.2.4 Sensors and devices Microscale sensors and microscale sensor-based devices/systems have already been used in construction and the built environment to monitor and/or control the environment conditions (e.g., temperature, moisture, gas/smoke, and noise) and the materials/structure performance (e.g., stress, strain, vibration, pressure, flow, corrosion, and deterioration). Nanotechnology approaches are not only enabling sensors/devices to be made much smaller, more reliable, and energy efficient but also opening up new possibilities beyond the reach of mere microscale manufacture (e.g., biomimetic sensors and CNT systems). For example, the piezoelectric effect of ceramic particles (lead-zirconate-titanate) incorporated into coating materials has been successfully exploited in the development of a piezoelectric paint which can be sprayed directly onto the surface of structures (e.g., bridges) to sense structural vibrations (Hale, 2001). The feasibility of Cyberliths, or Smart Aggregates, as wireless sensors embedded in concrete is being evaluated. In the future these microsensors might be reduced to dust particle size, with the ability to coat an entire bridge with Smart Dust for optimum monitoring capabilities via a smart sensor net. These sensors can be used remotely to monitor the condition of the concrete and reinforcement without damaging the structures (Balaguru, 2006).

Nanotechnology-based devices, nano-electromechanical systems (NEMS), can improve the understanding of the behaviour of concrete material during mixing and curing as well as when the concrete structure is put in service. For example, with wireless NEMS sensors, the density and viscosity of unset concrete in mixing and pumping equipment can be measured. Assembly of integrated systems and components at nanoscale for general industrial applications are also envisaged. Nanotechnology-enabled sensors/devices offer great potential for developing smart materials and buildings/structures with 'self-sensing' and 'self-actuating' capabilities, which thus are able to adapt to changing service and environmental or weather conditions.

3.4.1.3 Future prospect

Nearly a decade has elapsed since the start of the work of the EU NANOCONEX project and the first state-of-the-art report on nano-technology application in construction. Back then, nanotechnology was almost unknown outside the scientific circle; a Google search for nanotechnology and construction or buildings produced fewer than 10 results; and only very limited, isolated nanotechnology-related RTD activities existed in construction. Considering the very low starting point and the conservative nature of the construction indus-try, the progress made so far (as described in the previous sections) has been really remarkable. Though the relatively small number of practical applications still holds back much of the prospects for nanotechnology in construction and the built environment, the increasing emphasis on sustainability development and the push for knowledge-based economy on the construction sector is providing important new drivers for further progress in the field.

Overall, the outlook for nanotechnology application in con-struction is very favourable. Huge potential has been predicted for nanotechnology applications in construction, and even minor improvements in materials and processes could bring large accumu-lated benefits, considering the very large scale of activities involved in the construction industry.

Nanotechnology is now increasingly recognised as a route to achieve real competitive, sustainable growth and innovation in the construction industry by many researchers, industrial leaders, and government departments.

Bibliography

Andersen, M. M. and Molin, M. (2007). NanoByg - a survey of nanoinnovation in Danish construction. Available at http://www.risoe.dtu.dk/Risoe_dk/Home/Knowledge_base/publications/Reports/ris-r-1602.aspx.

Balaguru, P. N., Chong, K., and Larsen-Basse, J. (2006). Nano-concrete: possibilities and challenges, Proc. NICOM2, 233–243.

Bartos, P. J. M. (2009). Nanotechnology in construction: a roadmap for development, Proc. NICOM3, 15–26.

BASF. (2010). Lotus effect spray. Available at http://www.basf.de.

Bentz, D. P., Snyder, K. A., Cass, L. C., and Peltz, M. A. (2008). Doubling the service life of concrete structures, I: reducing ion mobility using nanoscale viscosity modifiers, Cement Concrete Composites, 30, 674–678.

Bioni, C. S. (2010). Available at http://www.bioni.de/index.php?page=produktprogramm&lang=en.

Castano, V. (2010). Deletum 5000 anti-graffiti paint. Available at http://www.nanotechproject.org/inventories/consumer/browse/products/4983/.

ChromoGenics. (2010). Smart windows. Available at http://www.chromogenics.com/index_eng.htm.

DTI. (2001). Constructing the Future, Foresight report, London.

Elvin, G. (2007). Nanotechnology for green building, Green Technol. *Forum*, 2007.

European Commission. (2010). EU directive on the energy performance of buildings. Available at http://ec.europa.eu/energy/efficiency/buildings/buildings_en.htm.

European Commission. (2010). EUR 24283 - Energy-efficient Buildings PPP, Multi-Annual Roadmap and Longer Term Strategy, 48.

Flanagan, R., Jewell, C., Larsson, B., and Sfeir, C. (2001). Vision 2020 – building Sweden's future. Available at http://www.bem.chalmers.se/sh/forskning/vision2020article.pdf.

Freedonia. (2007). Nanotechnology in construction – US industry study with forecasts for 2011, 2016 & 2025, 174, 2007. Available at http://www.reportlinker.com/p060851/Nanotechnology-in-Construction.html.

Gann, D. (2002). A Review of Nanotechnology and Its Potential Applications for Construction, SPRU (Science and Technology Policy Research), University of Sussex, October 2002.

Hale, J. (2001). Smart paint senses structural vibrations, Sensors Online, 18(8).

In Realis. (2002). A Critical Investor's Guide to Nanotechnology, February 2002.

Italcementi Group. (2010). Tx active cement. Available at http://www.italcementigroup.com/ENG.

Konsta-Gdoutos, M. S., Metaxa, Z. S., and Shah, S. P. (2010). Multi-scale mechanical and fracture characteristics and early-age strain capacity of high performance carbon nanotube/cement nanocomposites, Cement Concrete Composites, 32, 110–115.

Lotus-Effect®. (2010). Available at http://www.lotus-effekt.de/en/index.php.

Luther, W. and Buchmann, G. (2009). Nano.DE-Report 2009: Status Quo of Nanotechnology in Germany, Federal Ministry of Education and Research, 93.

Mann, S. (2006). Nanotechnology and Construction, Nanoforum Report, Nanoforum 2006. Available at www.nanoforum.org.

MMFX. (2010). Steel Corporation of America. Available at http://www.mmfxsteel.com/.

NanoPore Inc. (2010). Available at http://www.nanopore.com/vip.html.

Nansulate®. (2010). Available at http://www.nansulate.com/index.html.

Pilkington. (2010). Available at http://www.activglass.com/.

Porro, A. (2004). Nanotechnology, the way towards the materials of the future, Proc. B4E Conf. – Building Eur. Future, ECCREDI, Brussels, Belgium (October 2004), 1, 17–31.

Sanchez, F. and Sobolev, K. (2010). Nanotechnology in concrete – a review, *Construct. Building Mater.*, 24, 2060–2071.

Scalisi, F. (2009). Nanostructured materials in new and existing buildings: to improved performance and saving of energy, Proc. NICOM3, 351–356.

Zhu, W. (2007). Application of nanotechnology in construction materials, *Adv. Mater. Restoration*, 2, 77–90.

Zhu, W., Bartos, P. J. M., and Porro, A. (2004). Application of nanotechnology in construction-summary of a state-of-the-art report, *Mater. Struct.*, 37(11), 649–658.

3.4.2 *Active Windows for Daylight-Guiding Applications*

Andreas Jäkel, Qingdang Li, Jörg Clobes,
Volker Viereck, and Prof. Dr. Hartmut Hillmer
*Institute of Nanostructure Technologies and
Analytics and Center for Interdisciplinary
Nanostructure Science and Technology, Germany*

3.4.2.1 Introduction and basics

In the past few years, many applications could be improved by applying *micro-electromechanical systems* (MEMS) technology to common products. In the following sections, a system is presented that equips multi-glazed windows with an actuable micromirror system to enable light-guiding and light-blinding functionalities. These micromirrors can be produced with common thin-film and microsystem technologies and will be implemented in the windows themselves. A networked sensing system that controls the environmental conditions will provide information for the automated driving unit. The complete system will be a so-called *active window*.

Figure 3.33 depicts the simulation of a room equipped with active windows. While the left window remains in the opened state to illuminate the room, the micromirrors in the other windows have changed their position to shield the reading person from glare. In this case, light is guided to the ceiling to provide the reader just enough light to read his book. The light is dispersed at the ceiling to diffuse light, defining a comfortable, healthy, and bright ambience. Part of the right overhead window is also opened to give the plant on the table the required sunlight. The appearance of the window itself is comparable to a tunable tint. The view to the outside is unchanged contrary to most of the conventional blind systems.

When the reader now changes his position, the light possibly follows these movements. The networked sensing system detects this motion and readjusts the deflection angle of the micromirrors accordingly.

Thus, in a room equipped with an active window, the spot of light illuminating the surroundings of the person changes its position on the ceiling and follows the movements of that person.

Figure 3.33 Simulation of the working principle of the active window. The micromirrors guide the incoming daylight and protect the reading person from glare effects at the same time. See also Colour Insert.

Due to the improved utilisation of sunlight for the room's illumination, not only does the expendability of additional artificial light sources save energy, the active windows support air conditioning and heating installations and, thus, also lower the total energy consumption of the room. In some situations this simply can be done by just closing all the mirrors completely. In this state, the window becomes nearly 100% reflective in the visible as well as the infrared domain. Thus, in summer, unwanted thermal energy input can be avoided. In winter, the active window guides the thermal radiation to a wall opposite the window. This wall heats up and spreads heat energy stored little by little. The use of micromirrors for this application is not primarily driven by technology itself, but there are many chances of improvement compared with the original scope of operation of a window. In the same way, in many modern applications, microsystem technology is already applied to increase the functionality, contrary to conventional systems. Micromirrors or MEMS in general can be designed to have defined properties, well adopted for the requirements of the application. Due to their small size, MEMS hardly suffer from mechanical wear in contrast to most of the macroscopic mechanical systems (Li *et al.*, 2009). Their electrical and optical properties can be strongly influenced by design. In future, entire surfaces will be equipped with MEMS to give them functionalities beyond today's imagination.

Today, MEMS technology has already been implemented in many commercially available products (Wu, 2009). For instance accelerometres used to detect an accident for triggering the airbag in a car are produced that way (Monk, 2002). In particular, micromirrors already play an important role in commercially available products. They can be found in projectors like the digital light processing (DLP) beamers by Texas Instruments (Kessel *et al.*, 1998) and are proposed for car stereo displays (Pizzi *et al.*, 2003), for instance.

Figure 3.34 shows the schematic working principle of single micromirror elements designed for active windows. Encapsuled in between the two panes of a double-glazed, thermally insulating window, the single micromirror elements are protected against dirt and dust particles or other mechanical influences that could cause damages.

Their small size makes them almost insusceptible to wear, and the single micromirror elements cannot be resolved optically by the human eye any more. According to the design, segmentation makes them addressable as arrays of several thousand micromirror elements at a time. Fixed to the glass substrate (the first pane [1]) on just one side, each single element stays free to move. According to Figure 3.34 micromirror elements (5) are equipped with a metal electrode and the window pane (1) with a transparent one. By applying a voltage between those two electrodes, the attractive electrostatic force will pull the mirror to the glass pane. By design, the mirror elements are in their 'open position' without any voltage applied, that is, without actuation. This is caused by tailored mechanical stress in the mirror elements.

By applying a voltage, the micromirror elements are in equilibrium between electrostatic and restoring forces. Figure 3.34a shows the mirror's completely closed position, where the mirror

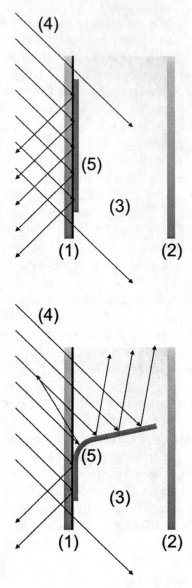

Figure 3.34 Schematic working principle of single micromirror elements inside a double-glazed window: a) in the fully actuated, completely closed state and b) deflected [6]; (1) first pane (in contact with the air outside the building), (2) second pane (in contact with the air inside the room), (3) noble gas isolation between the two panes, (4) schematic ray of sunlight, and (5) a single mirror. See also Colour Insert.

is located flat on the substrate. The window will become almost entirely reflective. Figure 3.34b depicts an opened mirror at a certain angle of deflection. By control of this angle, the light is guided to the desired position in the room. The micromirror elements remain in this defined position and change the path of light coming to pass the window until the voltage applied to the mirror elements is changed and another angle of deflection is appointed. An impression of a tinted window is the result.

Beside the electrostatic actuation, there is the possibility of magnetic (Miller *et al.*, 1995), thermal (Sinclair, 2000), mechanic (Fritz *et al.*, 2009), or electrodynamic (Urey, 2002) actuation. They are all less effective than electrostatic actuation (Legtenberg *et al.*, 1996). This is because magnetic actuation is hard to implement, since you need a large homogenous magnetic field to drive the micromirrors. For keeping an actuated state, a strong permanent current is required, revealing high power consumption.

Thermal actuation is suffering from unwanted heating of the device while the sunlight is heating it up, too. Some electrostatic actuations, for example, a combdrive actuator (Schenk, 2000), are much more precise than the one reported in our micromirror arrays; however, they need a lot of space for the mechanic part that is of a very big lateral size. Using the lateral one, we would reveal a lot of mechanics on the window pane but only few micromirrors to drive.

The single micromirror elements can be approximated as capacitors. Therefore, there are two possibilities, where they will consume energy themselves.

There is the actuation; each change of state will require an input of energy, but this is needed only at times. Also, there are leakage currents that cannot be avoided but can be minimised by design. These currents are very low – in the order of magnitude of only 0.1 W/m^2 in our case, at present. Since many of the present systems consume much more energy, while several systems have to be applied at the same time, there is a huge potential to save energy by the utilisation of active windows involving nanotechnology.

Systems are referred to as nanotechnological systems if at least one dimension is structured in a range below 100 nm. However, a more suitable definition is based rather on the variation of the geometry of structures than on their initial dimensions; material and device properties are hence not only determined by material choices but also by utilisation of nanoscale structuring. Thus, an improvement of device properties can be induced. Namely, optical and electrical properties can be tailored to the needs of the application desired, which makes it possible to implement, for example, waveguides inside a system. Additionally, the variation of mechanical properties strongly influences material fatigue and increases the process yield in the production process. Thus, it can improve system stability during operation and prolong its lifetime.

However, the ongoing miniaturisation has driven the MEMS technology into the nano dimension. For instance, state-of-the-art central processing units (CPUs) have structural sizes of several tens of nanometres. Therefore, terms like 'nano electro-mechanical systems' (NEMS) are gaining popularity at present.

Nevertheless, the active window system is inbetween the two; it has microscale dimensions in the lateral and nanoscale dimensions in the vertical direction. Therefore, it is referred to as MEMS here.

With the use of micromirror windows implemented into a complete active window, energy is saved by using daylight instead of artificial electrical light sources very efficiently. This is beneficial for health considerations because the human body is accommodated to the sun's light.

Many positive influences of natural daylight on the human body have been reported (Braun, 2008). Additionally, air-conditioning systems and heating installations can be supported by controlling the energy input through the manipilation of the incoming thermal radiation.

Thus, active windows based on nanotechnology save energy involving heating, air conditioning, and illumination.

Table 3.9 gives an overview of conventional systems in comparison to active windows. Our reported system can replace most of the conventional ones and minimise the required energy input.

3.4.2.2 Complete active window

Figure 3.35 shows a schematic of a room equipped with a complete active window system. The MEMS micromirror-modules are placed between the two panes of the thermally insulating window.

Table 3.9   Comparison of conventional systems and active windows

Device	Blind	Selectivity	Glare shield	Heat shield	View
Macroscopic light-guiding blinds	Reflection	Angle	Good	Good	Very good
Light shelfs	Reflection, light guiding	Angle	Poor	Poor	Very good
Light-guiding holograms	Diffraction	Angle, diffuse/direct light	Very good	Low	Very good
Reflecting blinds	Shadowing,	Angle	Very good	Flexible	Very good
Prismfoil light guides	Refraction	Angle	Very good	Poor	Poor
Solar glass	Reflection	Angle	Good	Very good	Very good
Light-guiding glass	Refraction	Angle	Good	Poor	Poor
Active windows	Reflection, light guiding	Angle	Very good	Very good	Flexible

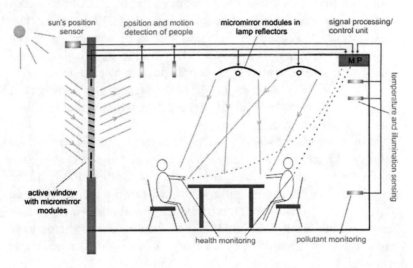

Figure 3.35 Schematic of a complete active window. The sensors gain information about the situation in the room, and the signal processor collects this information and provides it to the control unit, which drives the micromirror modules on the basis of these input parametres (Viereck *et al.*, 2008). See also Colour Insert.

A networked sensing system is monitoring the inside of this room and detects, for example, brightness distribution, temperature distribution, and the presence and health condition of people, as well as gathers information about the exterior environmental situation like the sun's position or the sky coverage, for instance. This way, for example, the sun's position or the brightest spot on the sky is monitored to utilise the micromirror array as efficiently as possible. Inside, the temperature and brightness as well as the position of people is detected to adjust the illumination situation according to the requirements. It will be possible to collect much more information. For instance, pollutant contamination or the health conditions of the people can be integrated into the networked sensing system. A signal processor collects all the information and provides it to the control unit that drives the MEMS modules inside the windows.

Guiding light is not necessarily the only function of the active window system. The scope of operation of active windows can be much wider. For example, if elderly people have fallen down on the floor or people have lost consciousness, this can be detected automatically by the networked sensing system. The system then can display the need of help by a specific switching of the micromirrors in the window itself. The window may display, 'I need help,' for instance, to alert other people passing by outside.

It can be used as a burglar alarm by checking its resistivity. A change of the resistance can display a defect window that has been probably broken by criminals.

Due to economic sustainability, the active windows will consist of several micromirror modules that are implemented next to each other in the window. Thus, the quality of the windows can be enhanced by reduction of pixel errors in the production process. These modules will be interconnected invisibly over the whole window pane. Each of these modules consists of several micromirror arrays of the size of approximately 1 dm^2 due to economic reasons.

Therefore, these arrays can be custom-made, according to the requirements of the application. Each change of size will be a trade-off between production cost and functionality. The micromirror arrays are formed of a manifold of single micromirrors that are electrically interconnected and, therefore, not adressable individually. Since the single mirrors are so small, this is not necessary even if the active window should work as a monochrome display by just switching very small arrays as pixels 'on' or 'off' (or grey scales). These modules are fabricated on substrates of glass or polymer foil.

Since single micromirror elements are that small, the human eye is not able to resolve these structures any more. Therefore, the view through the active window will remain unchanged. The invisibility of the system makes it possible to apply this technology to landmarked buildings as well. In those buildings it is forbidden to implement anything that will change the appearance of the facade, which makes it impossible to install conventional systems, as these are extremely visible.

3.4.2.3 Regulation concepts for active windows

To develop regulation concepts, an exemplary room has been simulated at different sun positions. Representative simulations of summer and winter scenarios are presented in Figure 3.36.

To demonstrate the beneficial effect of the active window, identical scenarios with a room with a regular window (left) and one equipped with an active window (right) can be seen. In summer, without an active window, the area next to the window is very bright, while the rest of the room stays rather dim. The active window guides most of the light to the ceiling. There, the light is dispersed, and a soft, diffused light illuminates the room. Now, people are able to work in a bright ambience without additional artificial light sources. In winter, the position of the sun is lower. In the room without the active window, people suffer from glare. With the active window, part of the light is guided to the ceiling to avoid glare. The entire light distribution of the room is improved.

The active window is a highly flexible system that is able to provide a wide range of functionalities that suits the special needs and requirements of every environmental setting.

This flexibility offers a high potential of architectural and individual creation, thus offering the possibility to set the ambience of every room to the optimum light situation. Due to its properties, it provides the possibility to save energy and improve the ambience of a room at the same time.

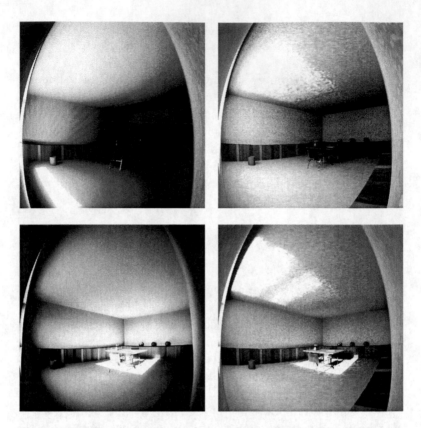

Figure 3.36 Simulations to develop regulation concepts. The 2×2 matrix shows the difference between windows with and without micromirrors for common sun positions at noon in summer and winter for a typical western European room at roughly 51 degrees of latitude (Viereck *et al.*, 2007). See also Colour Insert.

3.4.2.4 Production of micromirror arrays

The fabrication processes of micromirrors are based on microsystem and thin-film technologies using low-cost processes and materials. Many processes will occur unintroduced since their introduction lies beyond the framework of this section; thus, the interested reader is referred to Senturia (2000). A schematic process flow can be seen in Figure 3.37. All process steps are kept as simple as possible to avoid costly production. On the substrate (Li *et al.*, 2009), a transparent indium tin oxide (ITO) electrode and an electrically

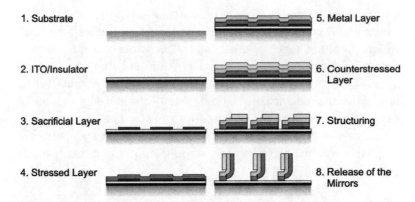

1. Substrate
2. ITO/Insulator
3. Sacrificial Layer
4. Stressed Layer
5. Metal Layer
6. Counterstressed Layer
7. Structuring
8. Release of the Mirrors

Figure 3.37 Schematic production process of the micromirrors. The material system is deposited (2–6) on the substrate (1), it is micromachined (7), and the mirrors are released (8) (Jäkel, 2009). See also Colour Insert.

insulating layer (Wu, 2009) are deposited. A so-called sacrificial layer (Monk, 2002) is structured. It is removed in the last process step (Sinclair, 2000) to release the micromirrors from the substrate so that they can self-deflect. Afterwards, the carrier layer (Kessel *et al.*, 1998) is deposited. It consists of a dielectric bilayer system and is mechanically pre-stressed to provide the upward self-deflection of the mirrors. A metal layer as a reflective mirror surface is applied (Pizzi *et al.*, 2003). To be able to steer the light, a quasi-planar surface of the mirrors is required. Therefore, a mechanically counter-stressed dielectric layer (Jäkel *et al.*, 2009) that avoids rolling-up of the whole mirror element is placed on top of the material system. Afterwards, the material system is microstructured by means of photolithography and etching processes (Miller *et al.*, 1995). In this step, the grid for the electrical interconnection of the micromirror arrays is formed, too. The single mirror elements are now ready to be released from the substrate (Sinclair, 2000).

The micromirror arrays are now ready for implementation into the multiglazed window. This process depends on the kind of substrate that is used. One possibility would be to use the window pane itself. The second way is the production on very flat glass substrates.

Another possibility is the production on polymer foil as a substrate. This way, the micromirrors can be produced on rolls, which simplifies logistics drastically. Afterwards, the interconnection of and the connection to the sensing and driving units are done. These steps can be integrated into the window-manufacturing process easily. The active window system offers a high degree of flexibility concerning the design. The sensing system, as well as the segmentation, offers a huge potential to adapt the properties of the active window according to the needs and requirements of the customer.

Bibliography

Braun, H. (2008). Photobiology: The Biological Impact of Sunlight on Health & Infection Control, Phoenix Project Foundation, 2008.

Fritz, J., Pinter, S., Friese, C., Pirk, T., and Seidel, H. (2009). Microscanner using self aligned vertical comb drives in a switched electrode configuration for large static rotation, IEEE/LEOS Int. Conf. Opt. MEMS Nanophoton., IEEE.

Jäkel, A., Li, O., Viereck, V., and Hillmer, H. (2009). Development of electrostatically driven micromirror arrays for daylight guiding on large areas, IEEE/LEOS Int. Conf. Opt. MEMS Nanophoton., IEEE, 41–42.

Legtenberg, R., Groeneveld, A. W., and Elvenspoek, M. (1996). Combdrive actuators for large displacements, *J. Micromech. Microeng.*, 6, 320–329.

Li, Q., Jäkel, A., Viereck, V., Schmid, J., and Hillmer, H. (2009). Design and fabrication of self-assembling micromirror arrays, SPIE Eur. Microtechnol. New Millennium, 736211.

Miller, R., Burr, G., Tai, Y. C., and Psaltis, D. (1995). Magnetically actuated micromirror use as optical deflectors, Proc. Electrochem. *Soc. Meeting*, 95–18, 474–480.

Monk, D. J. (2002). MEMS physical sensors for automotive applications, microfabricated systems and MEMS VI: proceedings of the international symposium by Peter J. Hesketh, *Electrochem. Soc. Meeting*, 6, 43–63.

Pizzi, M., Koniachkine, V., Nieri, M., Sinesi, S., and Perlo, P. (2003). Electrostatically driven film like modulators for display applications, *Microsyst. Technol.*, 10, 17–21.

Schenk, H. (2000). Ein neuartiger Mikroaktor zur ein- und zweidimensionalen Ablenkung von Licht. PhD thesis, Gerhard-Mercator-Universität-Gesamthochschule-Duisburg, Fachbereich Elektrotechnik, 2000.

Senturia, S. D. (2000). Microsystem Design, Kluwer Academic Publishers.

Sinclair, M. J. (2000). A high force low area MEMS thermal actuator, *Therm. Thermomech. Phenom. Electron. Syst.*, 1, 127–132.

Urey, H. (2002). Torsional MEMS scanner design for high-resolution display systems, *Opt. Scanning*, 4773, 27–37.

van Kessel, P. F., Hornbeck, L. J., Meier, R. E., and Douglass, M. R. (1998). A MEMS-based projection display, Proc. IEEE, 86(8).

Viereck, V., Ackermann, J., Li, Q., Jäkel, A., Schmid, J., and Hillmer, H. (2008). Sun glasses for buildings based on micro mirror arrays: technology, control by networked sensors and scaling potential, 5th Int. IEEE Conf. Networked Sensing Syst. (INSS), *Tech. Digest*, 135–139.

Viereck, V., Li, Q., Ackermann, J., Schwank, A., Araujo, S., Jäkel, A., Werner, S., Dharmarasu, N., Schmid, J., and Hillmer, H. (2007). Novel large area applications using optical MEMS, IEEE/LEOS Int. Conf. Opt. MEMS Nanophoton., IEEE, 55–56.

Wu, M. C. (2009). From O-MEMS to nO-MEMS, IEEE/LEOS Int. Conf. Opt. MEMS Nanophoton., IEEE.

3.4.3 Energy Efficiency Potential of Nanotechnology in Production Processes

Dr. Karl-Heinz Haas
Fraunhofer Institute for Silicate Research, Germany

3.4.3.1 Introduction

Nanomaterials play an important role in the application of nanotechnologies in the real world. They are the basic units of nanosystems, which interact with the macro world.

There are at least two different aspects of nanomaterials as far as **production** is concerned (Haas, 2003). Nanomaterials have to be produced – like other conventional materials also – and efficiency aspects (materials and energy efficiency) are of increasing importance. This is the **'production of'** aspect.

On the other hand, nanotechnologies and nanomaterials can be used in classical production processes to improve the quality and efficiency of processes. This is the '**producing with**' aspect. Both aspects will be covered within this contribution. Special emphasis will be given to energy efficiency aspects, wherever possible.

One has to bear in mind, however, that this is only one part of the story: life cycle aspects of nanomaterials and general resource efficiency have to be taken into account when judging eco-friendliness of nanomaterials and nanotechnologies.

The aspect of so-called molecular manufacturing will not be covered in detail in this contribution, since too many fundamental properties are not yet well understood.

3.4.3.2 Types and properties of nanoscaled materials

Nanomaterials come in different forms; they differ mainly in the confinement below 100 nm, starting with confinement in all three dimensions (particles 0D), in two dimensions (fibres, tubes 1D) to one dimension (thin films, multilayers 2D). A special case of 3D confinement are bulk nanostructured nanomaterials (see Figure 3.38).

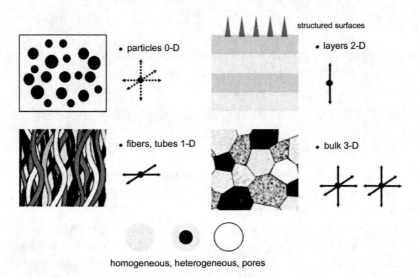

Figure 3.38 Dimensionality of nanomaterials. See also Colour Insert.

The properties of nanomaterials useful for new applications are mainly related to their high surface/interface areas, quantum effects, and biological principles (self-organisation, self-repair, adaptation, etc.).

The distinction between intrinsic and process nanomaterials is also very helpful (Haas, 2003). Intrinsic nanomaterials keep their nanostructure up to the final product, whereas process nanomaterials use the high reactivity of nanoscaled intermediates (particles, dispersions, etc.) but are not necessarily nanostructured in their final form. Examples of process nanomaterials are ceramic fibres processed, for example, by the sol-gel-method or metallic films based on nanoparticles.

3.4.3.3 Production processes of nanomaterials

The production processes used in the lab or in the industry can be categorised as top-down (starting with macroscopic materials) and bottom-up technologies (starting with atoms and small molecules), including techniques for nanostructuring materials and surfaces. Top-down methods are often used in microelectronics (lithography). The bottom-up techniques can further be differentiated by vapour-phase or liquid-phase technologies, including self-assembly.

Various techniques are shown in Table 3.10. Some of them are used on an industrial scale; some only on the lab scale due to their principal lack of scalability. Reviews about particle processing from the gas phase can be found in Kruis (1998), the principles of colloidal processing via the sol-gel technique are described in Brinker (1990). An assessment of manufacturing processes for nanomaterials was published by Lux research (Hartshorn, 2009). Large-volume synthesis for nanoparticles was described in Masala (2004). Several processing techniques like severe plastic deformation (SPD) and high-energy milling need high amounts of energy.

The nanoscale state of the materials has to be conserved up to the product level; therefore, dispergating nanoscale materials in matrices or keeping them stable in liquids is an essential requirement.

In situ processes for the synthesis of nanocomposites guarantee better dispersion since the connection between the nanoscale units is already preformed in the (inorganic-organic) precursor molecule

Table 3.10   Top-down and bottom-up nanomanufacturing methods based on Sengül (2008)

Top-down techniques	
Lithography	
Conventional lithography	Photolithography E-beam lithography (masc. generation)
Next generation lithography	Immersion lithography Extreme UV lithography X-ray lithography
Lithography with particles	E-beam lithography Focused ion beam lithography
Nanoimprint lithography	
Ethching	Wet etching
Dry etching	Reactive ion etching Plasma etching Sputtering
Electrospinning	
SPD (metals)	
Milling	Mechanical milling Cryo milling Mechanochemical milling/bonding
Bottom-up techniques	
Vapour-phase techniques	
Deposition techniques	Vapour-phase epitaxy CVD Molecular beam epitaxy Plasma-enhanced CVD Atomic layer deposition Sputtering Pulsed laser deposition
Nanoparticle/nanostructure synthesis techniques	Evaporation Laser ablation Flame synthesis Arc discharge
Liquid-phase techniques	Precipitation Sol-gel techniques Solvothermal synthesis Sonochemical synthesis Microwave irradiation Reverse micelle
Electrodeposition	
Self-assembly techniques	

(Haas, 2009). This can reduce energy needs for the dispersion process, but more chemistry is involved, which usually increases the price tag.

In the last few years, a general trend was observed to make chemistry more ecologically feasible and the term 'green chemistry/engineering' was coined (Dahl, 2007; Drzal, 2004; VDI, 2007).

The potential of nanotechnology to substitute hazardous substances was reported by Fiedeler (2008).

The general principles of green chemistry and engineering are shown in Table 3.11.

E-factor analysis, which summarises the material ressources needed in synthetic procedures, however, shows clearly that so

Table 3.11   Summary of the principles of green chemistry and green engineering (Eckerlman, 2008)

Principles of green chemistry	Principles of green engineering
Waste prevention	Inherent rather than circumstantial
Atom economy	Prevention rather than treatment
Less hazardous synthesis	Design for separation
Safer chemicals	Maximisation of mass, energy, space, & time efficieny
Safer solvents	Output pulled
Energy efficiency	Conservation of complexity
Renewables	Durability rather than immortality
Reduce derivatives	Minimisation of excess
Catalysis	Minimisation of material diversity
Design for degradation	Integration of material & energy flows
Real-time analysis	Design for afterlife
Inherently safer chemistry	Renewable rather than depleting

Table 3.12   Annual throughput and e-factor comparison of different material classes (Eckerlman, 2008)

Industry segment	Product throughput (log kg)	E-factor
Bulk chemicals	7 – 9	less than 1 up to 5
Fine chemicals	5 – 7	5 – 50
Pharmaceuticals	4 – 6	25 – 100
Nanomaterials	2 – 3	100 – 100,000

far, the processing of nanomaterials is not very efficient. The true possibilities of resource-effective synthesis for nanomaterials are far from being optimised. Especially purification and modification processes are not very resource efficient. New processing schemes have to be developed to unfold the true potential of nanomaterials for efficient green chemistry (see Table 3.12).

For selected nanomaterials, for example, CNTs, energy needs for different synthetic procedures have been evaluated together with the life cycle analysis of their composites (Euro, 2007; Kushnir, 2008). The possibilities to reduce carbon emissions by nanotechnology have been summarised in Cien (2007). More details about the environmental aspects of nanoproduction can be found in Section 4.2.

Another important aspect is that during production and processing, safety aspects, especially when dealing with free nanoparticles,
have to be well taken care of in order to ensure as little as possible
exposition of workers to airborne nanoparticles (Lauterwasser,
2005).

3.4.3.4 Nanotechnologies in production processes

3.4.3.4.1 Energy and manufacturing   The worldwide use of energy
is increasing steadily from around 500 EJ in 2005 to around 700 EJ
in 2030 (Nano, 2008). Currently, the industry consumes 30% of the
world's energy (Wec, 2004). The energy use scenarios indicate that
the industry sector worldwide will use 130 – 180 EJ per year in 2020
compared with 115 EJ in 2000. In 2050, the scenarios indicate 165 EJ
or 285 EJ, a 143% to 250% increase over today's usage. Energy
savings due to new-technology scenarios can be as much as 50 EJ
per year by 2020 and 120 EJ per year by 2050 (Wec, 2004; Seefeldt,
2007).

In Germany the energy for industry is around a quarter of the
total energy use and is predicted to decrease by about 7% until
2030 (Schulz, 2005). The most energy-intensive industries are metal
production and processing, thechemical industry, and the glass and
ceramic industries (UBA, 2009).

According to a Massachusetts Institute of Technology (MIT)
study in March 2009 (MIT, 2009), however, several modern
process technologies suffer from high inefficiencies. The 'study
sees alarming use of energy, materials in newer manufacturing
processes'. This was investigated for various industrial processes, as
shown below (Gutowski *et al.* 2009; Gutowski, 2004) in Figure 3.39.

The investigation did not cover all relevant industrial processes
and did not take into account air handling, etc., but it shows quite
clearly that modern processes are *highly energy intensive*.

It is generally accepted that energy efficiency is of increasing importance in various production processes (Luther, 2008;
BMBF, 2008; Almansa, 2006; Doe, 2009; NSF, 2002; NSF, 2007;
UBA, 2009; UK, 2006). The clean energy future using nanotechnology was described by Gillet (2002), Lux (2007), and Walsh,
(2007).

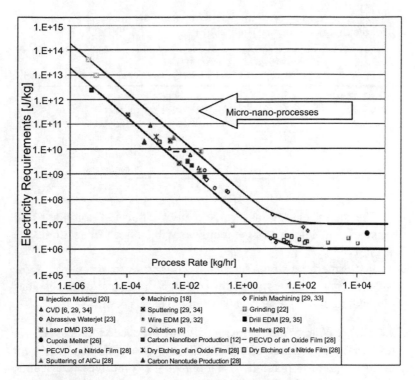

Figure 3.39 Work in the form of electricity used per unit of material processes for various manufacturing processes as a function of the rate of materials processing rate (Gutowski *et al.* 2009). Reprinted with permission from Gutowski, T., Branham, M., Dahmus, J., Jones, A., Thieriez, A., and Sekulic, D. (2009). Thermodynamic analysis of resources used in manufacturing processes, *Environ. Sci. Technol.*, 43, 1584–1590.

In the following section, it will be shown that selected nano-technologies can contribute to energy saving in various industrial processes.

3.4.3.4.2 Chemical processing   In 2006 a workshop paper was published, which gives detailed estimates for the energy-saving potential of chemical processing using nanomaterials and nano-technologies (Doe, 2006; Doe, 2007). The summary is shown in Table 3.13.

Table 3.13   Summary of estimated impact of nanomaterials by design for the chemical industry (Doe, 2006)

Segment	Cost savings ($ billion/year)	Energy savings (trillion BTU/year)	Nanomaterial application
Chemicals	2.5 – 4.0	200 – 400	Catalyst
Petroleum	0.2 – 0.8	80 – 200	Catalyst
Automobile	0.2 – 1.1		Catalyst
Shipping	2.5 – 3.4	150	Coatings
Manufacturing	1.8 – 3.5		Coatings
Natural gas	1.0 – 2.7		Membranes

The main energy savers are new catalysts, coatings, and membranes. Membranes technologies will be dealt in Section 3.2.3 of this book.

In Germany sustainability and energy were investigated for selected nano-applications in coatings, catalysis, and displays (Steinfeldt, 2004a) – see also Section 4.2 of this book.

3.4.3.4.3 Examples for industrial use in production processes   In this section, selected examples will be presented from the field of coatings, heat insulation, nanopowders, adhesives, lubricants, and heat-conductive nanofluids. Beside the examples described in this section, various other potentials of nanomaterials for energy efficiency may exist (NNI, 2004; Obse, 2009):

- energy-efficient production of concrete using nanomaterials,
- higher electrical conductivity using CNT cables or superconductors (Elcock, 2007), and
- CNT-reinforced ceramic or metal compounds: lightweight composites important for fast-moving parts (Samal, 2008).

An interesting technology analysis for the use of nanotechnology in production planning was published in 2008 (Heubach, 2008).

Coatings and surfaces

Hard coatings for tools: For generating high hardness of coatings, two nanotechnological principles can be used – multilayers and nanocomposites coatings. In the case of multilayers, the hardness is based on the ability of the system to slow down the movement of defects due to the high number of interfaces; therefore the high number of layers increases the hardness (Hauert, 2000).

The other possibility which is used for hard coatings is nanocomposite morphologies generated, for example, by modified physical vapour deposition (PVD) processes (Cselle, 2003; Cselle, 2009). For example, nanocrystalline TiAlN or AlCrN grains are embedded in an amorphous SiN matrix. The matrix prevents the growth of the grains and therefore guarantees the high hardness of the composite. Nanostructured hard coatings increase the cutting efficiencies of steel up to 50% (increase in cutting length or metres milled; Morstein, 2007).

Data for nanocoated toolings for various mechanical processing types can also be found in BalM (2009), BalS (2009), and Futu (2009).

Various nanocomposite coatings also show low-friction coefficients and show some anti-adhesive behaviour; therefore nanostructured coatings are also used in metal pressure foundries and lead to longer service time and less use of anti-adhesives (BalM, 2009). Also lubricant-free processing is possible in some cases.

Diamond-like carbon coatings (DLC): These coatings can have extremely high hardness and show low friction coefficients, which is important for a lot of tooling applications (Luther, 2009).

Transporting equipment: Also for polymer-based tools in production, enhancements in mechanical stability and better energy efficiency were described. For example, in paper-processing machinery, the surfaces can be modified with core-shell nanofillers in polymers, increasing the vibration-damping capability of the calander surfaces, leading to better mechanical performance at lower torque (Brew, 2008; Voit, 2007).

In another example transport bands were modified with silica nanofillers at a only 3 wt% in a TPU-isocyanate binder.

Even at this low loading friction coefficients could be lowered dramatically with 60% in the static case and 33% in the dynamic case (Falkenberg, 2006; Overmeyer, 2008).

Anti-scaling and corrosion protection: An impressive example is the use of non-permanent coatings on steel for better steel-shaping processes. A hybrid inorganic-organic polymer coating is applied on steel preforms before the high-temperature reshaping process, leading to a shielding against scaling. This is now used in the automotive industry (Sepeur, 2006; DE Patent Application 10-2007-038 214 A1). Corrosion protection is also of high relevance for industrial applications. In a study in 2004, the sustainability of hybrid inorganic-organic coating on aluminium which can be applied with coating thicknesses of only 5 microns was investigated (Steinfeldt, 2004a; Steinfeldt, 2004b). It was shown that the amount of varnishes and other materials could be greatly reduced (see also Section 4.2 of this book). The use of energy was also calculated. Also here the amount of energy was reduced, however, only to a lower extent.

Due to their low coating thickness, hybrid nanoscaled coatings can also be used as corrosion protection systems, for example, for Zn-rich, steel with the possibility of reshaping the coated parts without any cracking or loss of adhesion (Kron, 2007; Stumpp, 2009).

Easy-to-clean, low-energy coatings: Coatings with low surface energy are needed for various industrial processes. Nanoscaled coatings are already in use in different applications. Due to their hybrid inorganic-organic nature, they show good mechanical properties, which gives long service life to the coatings. For printing machinery various coatings have been described and are in use, for example, the nanochrom anti-sticking layer, which reduces the cleaning intervals for paper guide rolls (Jung, 2008; HMI, 2005). Another anti-soiling system for the printing industry was described by Weigt (2005), which is used by Heidelberger Print.

The modification of anti-stick polytetrafluorethylene (PTFE) materials with nano-alumina improves the mechanical wear resistance of the nanocomposites (McElwain, 2008; Sawyer, 2003).

In order to increase material efficiency, surfaces are also modified in order to reduce the interaction of filling goods with the container (Haas, 2007). Figure 3.40 shows an example of a low-energy surface for better removal of fluids.

A very impressive example for nanoscaled anti-adhesive coatings and their influence on energy efficiency are hybrid polymer coatings

Figure 3.40 Anti-wetting coating on plastic containers by plasma polymer treatment (Sängerlaub, 2008). See also Colour Insert.

on heat exchanger surfaces (Ritt, 2005; Ritt, 2009). Less dust is deposited on a hybrid polymer–coated metal heat exchanger surface. The coated surface stays cleaner for a much longer time, thus keeping the efficiency of heat transfer high and, therefore, decreasing the energy costs for the heat exchanger from 4,350 kWh to 2,940 KWh per year.

For medium-temperature applications, anti-stick coatings have also been developed, for example, for polymer-processing applications (Leykauf, 2004; Holocher, 2006). This system is used in duroplastic forming tools and increases the cleaning intervals dramatically (from 20 to 130 days).

In microelectronic processing the so-called lotus effect is now also used in order to keep patterns for flow paste processing cleaner for a longer time. This has been shown by Siemens recently (Siem, 2009).

For applications at higher temperatures in power plants or for metal foundries, anti-scaling coatings have been developed, for example, by ItN Nanovation AG (Meyer, 2006).

For metal foundries Nanocomp MP, a semi-permanent coating, is used as a lubricant in the casting process of aluminium, zinc, and brass with nanoscaled ZrO_2 as a binder (Hofmann, 2009). The production time before cleaning the forms could be increased from 300 parts to more than 750 parts. For the use in power plants, run-time Nanocomp PP was developed. It is based on BN as an anti-adhesive, high-temperature agent and is used in high-temperature heat exchangers to reduce cleaning times and to guarantee high heat transfer quality.

An application for the PV industry was described by Heraeus Company. In this case nanoscaled high-purity silica surfaces lead to better thermal processing at high temperatures (Hera, 2009).

Heat-insulating coatings: In gas turbine applications, the increase of efficiency of only 1% would save around 1 million euro per year (Luther, 2009) in a 400 MW unit. One possibility to increase the thermodynamic efficiency of gas turbines is to increase the working temperature. In order to achieve this, heat-resistant and heat-insulating materials are needed. Nanostructured ceramic coatings on metals can be generated by plasma-spraying on Ni base metals (Luther, 2009). Due to the reduced heat capacity of nanoscaled ceramics and the pores in the coatings, the heat conductivity is minimised.

An alternative heat-insulating system for medium temperatures uses nanoporous oxidic materials (SiO_2 based) with a binder. This is marketed under the trade name Nansulate. It could be shown in case studies that these coatings can improve energy efficiency (Nans, 2009).

A textile company claims 20% reduced energy costs when using this coating on burners, dye machines, steam boilers, and heat exchangers.

Porous systems, fibres, and webs

Heat insulation: Heat insulation is of increasing importance for various industrial and domestic applications. Due to their nanoscale

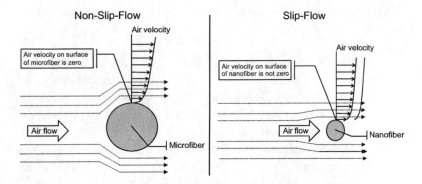

Figure 3.41 Air flow behaviour of micro- and nanofibres (Ruppertseder, 2008).

pores, various materials are interesting candidates for heat insulation since due to the confinement in the small pores, one basic heat transport mechanism (the collision of gas molecules with each other) is nearly eliminated. Nanofoams and aerogels are examples of these types of materials. Aerogels can be used as bulk materials or encapsulating webs up to temperatures of 600°C (Pyro, 2009).

Nanofibre webs: Polymer-based nanofibres can be processed by electrospinning with diameters down to only 20 nm (Plack, 2006). One main advantage of very thin fibres, besides their high surface area, is their reduced pressure drop in filtration due to the slip-flow behaviour, as shown in Figure 3.41.

This behaviour is used since in filtration webs for industrial applications. Nanowebs increase the removing capability for dust and other particles without sacrificing the air flow, therefore saving energy (Holl, 2004).

In a calculation of the Donaldson Company in the United States (Dona, 2008), power consumption using nanowebs can be lowered from 5 to 3 kWh at the same air flow compared with a conventional filter.

Joining processes: In industrial applications energy-efficient joining is of increasing interest. If the energy input for the joining process can be selectively controlled to only heat up the joining material, energy savings occur. Two examples are briefly described below.

- **Nanofoils**: Nanofoils are composed of separated reactive multilayers (e.g., Al and Ni) and can be used to join metal and ceramics (Wilden, 2008; Raic, 2008). The foils react chemically and generate enough heat to melt the solder but do not heat the ceramic or metal components significantly.
- **Adhesives with magnetic nanoparticles – bond and disbond on-command**: A new possibility for joining processes is the use of magnetic nanoparticles in the adhesive layer.

Due to their small particle size, magnetic nanoparticles can transfer the magnetic energy into thermal energy. The energy input can be controlled by electromagnetic energy (microwaves) so that only the adhesive is cured but there is little or no heating up of the surroundings. After the product life, the joined parts can be disbonded using higher microwave energies that destroy the adhesive layer (Sust, 2003).

Nanoparticles and dispersions

Sintering of nanoscale metallic and ceramic particles: One main reason for the use of nanoparticles in the processing, for example, of ceramics and metals, is their high surface energy, which can lead to a reduction of the processing temperature during sintering. The mechanical properties of nanocrystalline materials have been reviewed in 2006 (Memi, 2006). The particle size dependency of the sintering temperature for various metals can be found in (Wilden, 2006).

It can be noted that the reducing of the sintering temperature already begins well above the nanoregime. Nanoparticulate carbide powders are often used for tooling applications (Brookes, 2008). The effect of the enhanced mechanical properties of WC-nanoscale materials was described in IWGN (1999).

The behaviour of ceramic nano- and microparticles sintering was described in Ewsuk (2003). It was found that the sintering density using nanoscale ZnO can be reached at much lower temperatures compared with the micropowder.

An example of the advantage of nanoscaled ceramic powders is multilayers for electroceramic applications like gas sensors. Bosch presented a multilayer ceramic high-temperature oxygen sensor in 2004 (European Patent 0733204 B1). The sintering temperature can be reduced to below 1,200°C so that temperature-sensitive functional ceramics (perowskites) can be integrated.

When processing bulk ceramic or metal parts based on nanoscale powders, however, the sintering technique has to be adapted. Fast grain growth should be avoided; therefore fast sintering techniques are necessary (Seal, 2004).

Silver nanoparticles and nano-inks: One special case for the application of metallic nanoparticles is silver nano-inks. Due to their high surface energy, they can be processed and sintered at reduced temperatures, which makes them interesting materials for conductive layers in polymer electronics even on plastic substrates (Cabo, 2008). They can be processed by inkjet technology, a very fast process for generating structures on surfaces (Perelaer, 2007).

Controlling of viscosity of polymers melts: In polymer-processing technology, the melt viscosity is of great importance. It determines the temperature need of the proper melt flow and the fineness of the structures possible. Reducing the melt viscosity without changing any other advantageous properties is therefore very interesting. This was shown by the BASF some years ago (BASF, 2005).

By adding minor amounts of nanoscaled additives, the organization could show that melt viscosity decreased in the case of polybutylene terephthalate (PBT) (Weiss, 2004; Eipper, 2006). The mechanical properties of PBT remained unchanged, and energy savings of 20% could be realised (Leptich, 2005).

Improved heat transfer fluids: Heating and cooling fluids for industrial applications often show a small heat transfer coefficient due to the small heat conductivity of the liquids used (Wang, 2008; Wong, 2010). Since nanoparticles – properly dispersed – are stable against sedimentation, adding nanoparticles to the process fluids can increase the thermal conductivity, which is a good basis for improving the energy flow in industrial process and, therefore, finally, save energy. An overview about possible thermal conductivity enhancements can be found in Shen (2007). Multiwalled CNTs show improved thermal conductivities in synthetic oil of up to 150%.

Chemical mechanical polishing with nanoparticles: For many years, the processing of multilayers in microelectronics is improved by nanoparticles for chemical mechanical polishing in order to generate ultra-flat surfaces which are very important for multilayer microelectronics. Every defect in a layer which cannot be removed during processing will be also seen in the following layers and, therefore, will limit the function of the system (Zantye, 2004; Philipossian, 2003). Nanoparticles (SiO_2, CeO_2, Al_2O_3, etc.) are used due to their small size (removing also very small defects without creating new ones) and the possibility to modify their morphology (spherical, multi-facetted).

Nanolubricants: Tribology is a key factor in a lot of industrial processes (Martin, 2008). Whenever possible the friction should be minimised in order to save mechanical energy. Beside tribological tool coatings (DLC, etc.), nanolubricant additives are also used as additives for oil (Cant, 2009).

Sensors for process control and optimisation: Process control is also an important parameter in order to save energy in production processes. Sensors have been used in various manufacturing processes for quite some time. The development of nano-enhanced sensors (higher sensitivity, microintegration) could also lead to a contribution to energy efficiency in production processes (Robb, 2007; Stetter, 2006).

Examples of nano-enhanced sensors are nanoporous metal oxides, CNT-based sensors, and sensors based on the so-called giant magneto-resistance (GMR) effect. GMR sensors show high sensitivity for fast position detection in micromechanical systems (Luther, 2009). An example of the use of Ag-nanoparticle-based sensors is described in Marks (2002).

3.4.3.5 The vision of molecular manufacturing

Starting with the famous speech of Feynman in 1959, the idea of building things 'atom-by-atom', the so-called molecular nanotechnology (Wiki, 2007) came up at least on a theoretical basis. If this new processing would become real, production and manufacturing would be of course very much different from what we know now (Balzani, 2007). Also the mechanical tools would operate on a molecular basis.

Figure 3.42 shows molecular units (rotating gears) based on modified CNTs in a computer simulation.

The gears rotate under the action of a laser at six trillion rotations per minute (calculated) and have to be cooled by helium.

The assembly of macroscopic structures – will still be needed since people live in the macroworld – would be a huge task. The molecules could be synthesised by chemistry and then have to self-assemble in order to have highly parallel processes.

The possibility of molecular manufacturing is part of an ongoing discussion of theoreticians, visionaries, and chemists. Some basic problems are still not solved. What about the extreme surface energies of single nano-units? How can they be handled ('sticky finger' problem)? Where does the energy for the nanomachine come from? Nano-units are highly mobile, so how can we generate stable products for the macroworld?

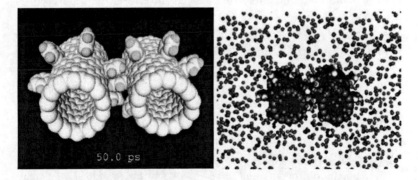

Figure 3.42 Molecular dynamic simulation of nanotube rotating gears based on functionalised CNTs (Han, 1996). See also Colour Insert.

The visions will be there for quite some time, and maybe new insight coming from the lab will realise these visions or sweep them away (Ball, 2010; Boeing, 2007).

We could and will learn more from nature since nature is the true master of self-organisation of complex systems in biosystems.

3.4.3.6 Conclusion, summary, and outlook

As shown in this contribution, nanotechnology has different aspects as far as energy efficiency is concerned.

Nanomaterials have to be manufactured, and this is so far a process with high energy needs.

On the other hand, nanotechnology can help to decrease the energy needs in various production processes. Very often, however, the total energy balance for the use of nanomaterials is not properly calculated. Can the energy saving by using nanomaterials in production processes outbalance the increased energy use for the synthesis of nanomaterials? In principle the atomic precise manufacturing should lead to the highest materials efficiency possible (Fore, 2007), but at what energy costs? In conclusion: Tremendous improvements are needed in production processes and equipment until nanomaterials can really be part of a 'green nanotechnology'. However, first, interesting applications in energy efficiency in manufacturing are here, and there will be more. So let us look at the total picture, which is process sustainability. This includes material and energy resources and cannot be separated.

Maybe something from the energy efficiency in transport can be learned for production processes. Why not save energy by temporarily storing energy also in production processes, maybe even with nanomaterials?

Bibliography

Almansa, A., Wögerer, C., Rempp, H., and Gebauer, M. (2006). IPMMAN: roadmap for micro- and manomanufacturing, version 2.0 August 10th 2006. Available at http://www.minamwebportal.eu/downloads/roadmaps/2007-06-25%20-%20Roadmap%20IPMMAN.pdf.

Ball, (2010). Welcome to the machine, *Chem. World*, January 2010, 56–60.

BalM. (2009). Balinit erhöht die Ausdauer beim Metalldruckgiessen, Brochure Oerlikon Balzers, 2009 (in German).

BalS. (2009). Balinit garantiert Bestform beim Stanzen und Umformen, Brochure Oerlikon Balzers, 2009 (in German).

Balzani, V., Credi, A., and Venturi, M. (2007). Molecular devices and machines, *Nano Today*, 2, 18–25.

BASF. (2005). Ultradur High-Speed, Folder Product Description BASF AG.

BMBF. (2008). Energieeffizienz in der Produktion - Untersuchung zum Handlungs- undForschungsbedarf, Studie im Auftrag des BMBF Juni 2007–Februar 2008 (in German).

Boeing, N. (2007). Abschied von der Nanovision, Die Zeit, August 2, 2007.

Brew, D. and Leithner-Kuzmany, T. (2008). Neue Walzenbezüge im Finishingbereich, TwoGether VOITH Paper-J., 25, 64–65.

Brinker, C. J. and Scherer, G. W. (1990). Sol-Gel Science - The Physics and Chemistry of Sol-Gel Processing, Academic Press, San Diego.

Brookes, K. (2008). Boosting hardmetals by using double carbides in the mix, Metal Powder Net MPR, May 2008, 10–18.

Cabo. (2008). Data sheet: Cabot conductive ink CCI-300, *Brochure Cabot Printed Electronic Materials*, 2008.

Canter, N. (2009). EP nanoparticles-based lubricant package, *Tribol. Lubrications Technol.*, 12–13.

Cien. (2007). Nanotech and cleantech - reducing carbon emissions today, *White paper Cientifica*, March 2007.

Cselle, T., Morstein, M., Coddet, O., Geisser, L., Holubar, Jilek, M., Sima, M., and Janak, M. (2003). Nanostructured coatings for high performance tools, Werkzeugtechnik, 77, 1–7.

Cselle, T., Büchel, C., Coddet, A., Lümkemann, A., Morstein, M., Moschko, A., and Prochazka, J. (2009). Flexible Beschichtung von TiN über Nanocomposite und Oxide zu DLC, Industrie-Workshop "PVD-Beschichtung von Werkzeugen und Bauteilen" Schmalkalden 16.6.2009. Available at www.platit.com (in German).

Dahl, J. A. (2007). Synthesis of functional nanomaterials within a green chemistry context, Dissertation, University of Oregon, December 2007.

DOE. (2006). Los Alamos National Lab estimated energy savings and financial impacts of nanomaterials by design and selected applications in the chemical industry, Vision 2020 Chemical Industry of the future roadmap.

DOE. (2007). Nanomanufacturing for energy efficiency, Report December 2007, Workshop Department of Energy DOE.

DOE. (2009). Energy matters: energy efficiency & renewable energy, Report US Department of Energy DOE, Fall 2009, 1.

Dona. (2008). Ultra-web: high efficiency nanofiber filters, Brochure Donaldson Filtration Solutions.

Drzal, L. T. (2004). Nanotechnology applications for green manufacturing, U.S. EPA 2004 Nanotechnology Science to Achieve Results (STAR). Progress Review Workshop – Nanotechnology and the Environment II, August 18–20, 2004.

Eckerlman, M., Zimmerman, J., and Anastas (2008). Towards green nano: E-factor analysis of several nanomaterial syntheses, *J. Ind. Ecol.*, 12, 316–328.

Eipper, A. and Völkel, M. (2006). Producing intricate parts economically, Kunststoffe, 11, 129–132.

Elcock, D. (2007). Potential impact of nanotechnology on energy transmission applications and needs, Rep. Argonne Natl. Lab. ANL/EVS/TM/08-3, November 2007.

Ewsuk, K., Diantonia, C., Ellerby, D., and Bencoe, D. (2003). Application of master sintering curve theory to predict and control nano-crystalline ceramic sintering, Sintering 2003, 3rd Int. Conf. Sci., Technol. Appl. Sintering, September 14–17, University Park, PA.

Falkenberg, S. and Wennekamp, F. (2006). Ganz klein für ganz gross – Nanopartikel im Transportband, phi Produktionstechnik Hannover, 7, 12–13 (in German).

Fiedeler, U. (2008). Using nanotechnology for the substitution of hazardous chemical substances, *J. Ind. Ecol.*, 12, 307–314.

Fore. (2007). Productive nanosystems - a technology roadmap, Batelle Memorial Institute, Foresight Nanotech Institute, October 2007, 1–198.

Futu. (2009). Balinit futura nano: the reliable solution for greater profitability, Broschure Oerlikon Balzers, 2009.

Gillett, S. L. (2002). Nanotechnology: clean energy and resources for the future, White paper, Foresight Institute.

Gutowski, T. (2004). Design and manufacturing for the environment, in Preprint Handbook of Mechanical Engineering, Springer Verlag.

Gutowski, T., Branham, M., Dahmus, J., Jones, A., Thieriez, A., and Sekulic, D. (2009). Thermodynamic analysis of resources used in manufacturing processes, *Environ. Sci. Technol.*, 43, 1584–1590.

Haas, K.-H., Hutter, F., Warnke, and Wengel, J. (2003). Abschlussbericht: Produktion von und mit Nanomaterialien - Untersuchung des F&H Bedarfes für die industrielle Produktion, BMBF-Projektträger Produktion und Fertigungstechnologien – Forschungszentrum Karlsruhe, BMBF Förderkennzeichen 02PH2107 2003 (in German).

Haas, K.-H. and Heubach, D. (2007). NanoProduktion: Innovationspotenziale für hessische Unternehmen durch Nanotechnologien in Produktionsprozessen Band 6 Schriftenreihe Aktionslinie Hessen Nanotech August 2007 (in German).

Han, J., Globus, A., Jaffe, R., and Deardorff, G. (1996). Molecular dynamics simulation of carbon nanotube based gears, NASA Techreport 1996. Available at http://alglobus.net/NASAwork/papers/MGMS_EC1/simulation/paper.html.

Hartshorn, C., Pekarskaya, E., Schiamber, B., Brown, I., and Udupa, S. (2009). Assessment of manufacturing processes for nanomaterials and nanointermediates, Lux Research Statement of findings, January 2009, 59 pages.

Hauert, R. and Patscheider. (2000). From alloying to nanocomposites - improved performance of hard coatings, *Adv. Eng. Mat.*, 2, 247–259.

Hera. (2009). Photovoltaik-Industrie profitiert von Heraeus Produkten, Press communication Fa. Heraeus 15.9.2009 (in German).

Heubach, D. (2008). Eine funktionsbasierte Analyse der Technologierelevanz von Nanotechnologie in der Produktplanung, Dissertation Universität Stuttgart (in German).

HMI. (2005). Nanochrome: Hohe Einsparungen für industrielle Anwendungen durch nanotechnologisch funktionalisierte Chromoberflächen, Presseinformation Hannovermesse 2005 Nanogate/Hartchrom AG (in German).

Hofmann, V., Meyer, F. Faber, S., and Bauch, H. (2009). Nanokeramische Beschichtungen steigern Effizienz und Qualität im Al-Kokillenguss, Giesserei, 96, 32–37 (in German).

Holl. (2004). Nanofaserbeschichtung für Filterwerkstoffe - Fa. Hollingsworth & Vose, Dokumentation Hessischer Innovationspreis 2004, 12–13 (in German).

Holocher-Ertl, M. and Passler, T. (2006). Nanokomposite-Materialien für die Beschichtung von Werkzeugen, GAK - Gummi, Fasern, Kunststoffe, 59, 91–95 (in German).

IWGN. (1999). IWGN 1999 nanotechnology research directions - IWGN workshop report: vision for nanotechnology research and development in the next decade, IWGN - Report National Science and Technology Council NSTC - WTEC, Loyola College of Maryland, September 1999.

Jung, M. (2008). Nanosystems development using functional nanomaterials, Rusnanotech Int. Forum Moskau, December 3–5, 2008.

Kron, J. (2007). Chrom(VI).-substitution: Umweltverträglicher Korrosionsschutz durch hybride Nanomaterialien, Ann. Rep. Fraunhofer-ISC 2007, 24–25 (in German).

Kruis, F., Fissan, H., and Peled, A. (1998). Synthesis of nanoparticles in the gas phase for electronic, optical and magnetic applications - a review, *J. Aerosol Sci.*, 29, 511–535.

Kushnir, D. and Sanden, B. (2008). Energy requirements for carbon nanoparticle production, *J. Ind. Ecol.*, 12, 360–375.

Lauterwasser, Ch. (2005). Small sizes that matter: opportunities and risks of nanotechnologies Allianz: report in co-operation with the OECD International Futures Programme, 1–44.

Leptich, J. (2005). Label Eco-Efficiency Analysis Ultradur High Speed BASF AG, December 2005.

Leykauf, M. (2004). Nanotechnik erhöht Standzeit von Werkzeugformen KE Komponenten und Systeme, 12, 56–57 (in German).

Luther, W. (2008). Einsatz von Nanotechnologien im Energiesektor, Band 9 Schriftenreihe Aktionslinie Hessen Nanotech, Mai 2008.

Luther, W. and Bachmann, G. (2009). Nano.de Report 2009: Status Quo der Nanotechnologie in Deutschland, BMBF-Broschüre erstellt von VDI-Technologiezentrum Abt. Zukünftige Technologien (in German).

Lux. (2007). Nanotech impact on energy and environmental technologies Lux Research, June 2007.

Marks, L. (2002). Transportation Nanotechnology Presentation - Center for Nanotransportation Nanotechnology 2002.

Martin, J. M. and Ohmae, N. (2008). Colloidal Lubrication: General Principles Nanolubricants, John Wiley & Sons, 1–13.

Masala, O. and Seshadri, R. (2004). Synthesis routes for large volumes of nanoparticles, *Ann. Rev. Mater. Res.*, 34, 41–81.

McElwain, S., Blanchet, A., Schadler, L., and Sawyer, W. G. (2008). Effect of particle size on the wear resistance of alumina-filled PTFE, *Micro Nanocompos. Tribol. Trans.*, 51, 247–253.

Meyer, F., Faber, S., and Novy, A. (2006). Nanocomp PP: Nanokeramische Schutzschichten für Kraftwerke im Praxistest, VGB Powertech Int. J. Electricity Heat Generation, 9, 131–135 (in German).

Meyers, M. A., Mishra, A., and Benson, D. J. (2006). Mechanical properties of nanocrystalline materials, *Prog. Mat. Sci.*, 51, 427–556.

MIT. (2009). Manufacturing inefficiency: study sees alarming use of energy, materials in newer manufacturing processes, MIT PHYSorg.com, March 18, 2009.

Morstein, M., Cselle, T., and Coddet, O. (2007). Senkung der Produktionskosten durch Verwendung von Nanocomposite-Schichten, Vortrag 7. Swissmem Zerspanungsseminar Winterthur und Olten, 23/24.07. 2007 (in German).

Nano. (2008). Cleanteach as a growth market: protecting resources with nanotechnology, Nanoage – Mag. Nanostart AG, 04/08 2008.

Nans. (2009). Nanosulate advanced industrial products, Case studies, Industrial Nanotech Inc., June 30, 2009.

NNI. (2004). Nanoscience research for energy needs, Report of the National Nanotechnology Initiative Grand Challenge Workshop, March 16–18, 2004.

NSF. (2002). Workshop on nanomanufacturing and processing, NSF-EC Workshop Puerto Rico, January 5–7, 2002.

NSF. (2007). Manufacturing at the nanoscale, Report of the National Nanotechnology Initiative USA Workshops, 2002–2004, 2007.

Obse. (2009). Economic analysis of nanotechnology for energy applications, Report EU-ObservatoryNano-project, 17.6.2009, 1–34.

Overmeyer, L. and Falkenberg, S. (2008). Dotierung von Transportbandmaterialien mit nanoskaligen Füllstoffen zur Optimierung der tribologischen Eigenschaften, Präsentation BMBF-Projekt DotTrans, Mai 2008 (in German).

Perelaer, J., Smiths, J., Hendriks, C. E., van Osfach, T. H., and Schubert, U. S. (2007). Low temperature sintering of conductive silver tracks, Presentation DPI-HTE Inkjet Workshop, June 2007, Eindhoven.

Philipossian, A. (2003). Neue Partikel für das chemisch-mechanische Planarisieren integrierter Schaltkreise, Degussa Science Newslett., 5, 30–33 (in German).

Pyro. (2009). Pyrogel XT: Flexible Industrial Insulation for High-Temperature Applications, Brochure Aspen Aerogels.

Raic, K., Rudolf, R., Todorovic, A., and Anzel, I. (2008). Multilayered nano-foils for low temperature metal-ceramic joining, Metall. - J. Metall., 14, 143–154.

Ritt. (2005). Nanotechnologie lässt Schmutz keine Chance - Innovative Schutzbeschichtung für Rittal TopTherm-Kühlgeräte, Company brochure Fa. Rittal, November 2005 (in German).

Ritt. (2009). Energie-Effizienz in Kühlsystemen, Company Brochure Fa. Rittal, Januar 2009 (in German).

Robb, D. (2007). Applying nanotechnology: submicron materials advance turbine design and operation, Turbinemachinery Int., July/August 2007, 39–41.

Ruppertseder, W. (2008). Ertl Th. Innovative filter media by means of integrated nanofilters, *Tech. Text.*, 51, E115.

Samal, S. S. and Bal, S. (2008). Carbon nanotube reinforced ceramic matrix composites - a review, *J. Miner. Mater. Charact. Eng.*, 7, 355370–.

Sängerlaub, S. (2008). Perspektiven der Nanotechnologie für verpackte Lebensmittel Vortrag BÖLW Fachtag: Nanotechnologie in der Lebensmittelwirtschaft Berlin, Juni 27, 2008 (in German).

Sawyer, G., Freudenberg, K., Bhimaraj, and Schadler, L. (2003). A study on the friction and wear behaviour of PTFE filled with alumina nanoparticles, Wear, 254, 573–580.

Schneider, R. (2009). Vortrag Fa. Rittal Nanotechnologie in der Oberflächenbehandlung Seminar für Metall bearbeitende Betriebe Mittwoch, November 4, 2009, Haus der Wirtschaft, Stuttgart (in German).

Schulz, W. (2005). Kurzbericht: Die Entwicklung der Energiemärkte bis zum Jahr 2030, Studie im Auftrag des BMWi, Mai 2005, Dokument Nr. 545 (in German).

Seal, S., Kuiry, S., Georgievea, and Agarwal, A. (2004). Manufacturing nanocomposite parts: present status and future challenges, MRS Bulletin, 29, 16–21.

Seefeldt, F., Wünsch, M., and Matthes, U. (2007). Potenziale für Energieeinsparung und Energieeffizienz im Lichte aktueller Preisentwicklungen, Final report BMWi-Projekt 18/06, August 2007 (in German).

Sengül, H., Theis Th. L., and Ghosh, S. (2008). Towards sustainable nanoproducts: an overview of nanomanufacturing methods, *J. Ind. Ecol.*, 12, 329–359.

Sepeur, S. (2006). The company Nano-X GmbH: Products for the automotive industry, Presentation Deutsche Börse AG Frankfurt, July 10, 2006.

Shen, B. (2007). Minimum quantity lubrication grinding using nanofluids, Dissertation University of Michigan, 2007.

Siem. (2009). Nanotechnik macht es möglich: Lotuseffekt für Leiterplatten, Elektronik-J., 9a, 7 (in German).

Steinfeldt, M., Petschow, U., Haum, R., and Gleich, A. V. (2004a). Nanotechnology and sustainability, Discussion paper of the IOEW 65/04 2004.

Steinfeldt, M., Gleich, A. V., Petschow, U., Haum, R., Chudoba Th., and Haubold, S. (2004b). Nachhaltigkeitseffekte durch Herstellung und Anwendung nanotechnologischer Produkte, BMBF-Studie: Schriftenreihe IÖW 177/04 Berlin, November 2004 (in German).

Stetter, J., Hesketh, and Hunter, G. (2006). Sensors: engineering structures and materials from micro to nano, Electrochem. Soc. Interface, Spring 2006, 66–69.

Stumpp, B. (2009). Gut gewappnet: Ormocer-Schichten schützen und bieten volle Umformbarkeit, KE Konstrukt. Eng., April 2009, 80–81 (in German).

Sust. (2003). Sustech GmbH Klebstoff mit Nano-Antennen, Brochure Konzeptionen für Nachhaltiges Wirtschaften BMBF-DLR-Umwelttechnik 2003 (in German).

UBA. (2009). Umweltwirtschaftsbericht 2009, Umweltbundesamt und Bundesministerium für Umwelt, Naturschutz und Reaktorsicherheit 2009 (in German).

UK. (2006). Sustainable production and consumption, UK technology strategy: Key technology area.

VDI. (2007). Nanotechnology in industrial applications, Proc. EuroNanoforum 2007 Düsseldorf, Juni 19–21 2007.

Voit. (2007). Fa. Voith NanoPearl: Die neue Premium-Generation der Kalanderbezüge, Company brochure Fa. Voith, August 2007 (in German).

Walsh, B. (2007). Environmentally beneficial nanotechnologies: barriers and opportunities, Oakden Hollins Report for Department for Environment, Food and Rural Affairs UK, including appendix 2007.

Wang, X.-W. and Mujumdar, A. S. (2008). A review on nanofluids - part II: experiments and applications, Braz. *J. Chem. Eng.*, 25, 631–648.

WEC. (2004). Energy end-use technologies for the 21st century, Report World Energy Council 2004, 1–140.

Weigel. (2009). Weigel, Hybride Nanomaterialien schützen vor Korrosion, Annual Report Fraunhofer-IWU 2008–2009, 29 (in German).

Weigt, W. and Herzbach, L. (2005). Nanobased functional surfaces in sheetfed offset printing presses, IIR-Proc. Nanotrends 2005, München, June 2005.

Weiß, C. and Eipper, A. (2004). Nano sorgt für Speed, Plastverarbeiter, 55, 178–179 (in German).

Wiki. (2007). Molecular Nanotechnology, Wikipedia, 8.11.2007.

Wilden, J., Bergmann, J., Schlichting, S., and Bruns, C. (2006). Nanotechnologie: Innovationspotentiale und Herausforderungen für die Fügetechnik, Abschlussbericht DVS Forschungsseminar 26.1.2006 (in German).

Wilden, J. (2008). Neuartige Fügetechniken durch Ausnutzen von Grösseneffekten, Mitteilungen aus dem Produktionstechnischen Zentrum Berlin FUTUR 1/2008 12–13 (in German).

Wong, K. V. and De Leon, O. (2010). Applications of nanofluids: current and future, *Adv. Mech. Eng.*, 11 pages, doi: 10.1155/2010/519659

Zantye, Kumar, A. and Sikder, A. K. (2004). Chemical mechanical planarization for microelectronics applications, *Mater. Sci. Eng.* R45, C6, 89–220.

Chapter 4

Potential Analysis and Assessment of the Impact of Nanotechnology on the Energy Sector Until 2030

4.1 Methodological Approach

Jochen Lambauer, Dr. Ulrich Fahl, and
Prof. Dr. Alfred Voß
*Institut für Energiewirtschaft
und Rationelle Energieanwendung (IER),
Universität Stuttgart, Germany*

In the scope of this book, the direct and indirect impacts of current and future products, production processes, and services of nanotechnology on the energy sector are analysed and evaluated. Chapter 4.2 gives an overview of the environmental impact and energy demand of nanotechnology. The potential analysis begins with the evaluation of the possible impacts on a changing energy demand as well as the required energy sources by identifying the status quo and the possible direct and indirect impacts of nanotechnology on all branches of industry, i.e., the change in energy demand as well as reduction potentials and possible efficiency enhancements. Indirect

Nanotechnology and Energy: Science, Promises, and Limits
Jochen Lambauer, Ulrich Fahl, and Alfred Voß
Copyright © 2013 Pan Stanford Publishing Pte. Ltd.
ISBN 978-981-4310-81-9 (Hardcover), 978-981-4364-06-5 (eBook)
www.panstanford.com

influences refer to the use of products, applications and services based on nanotechnologies (see Figure 4.1).

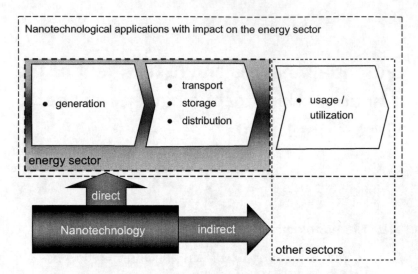

Figure 4.1 Direct and indirect impacts of nanotechnology.

The analysis includes a detailed examination of the energy sector with regard to innovations and improvements based on nanotechnology in order to envisage the possible future developments and changes.

Subsequently, it is possible to describe the impacts of selected nanotechnological applications on energy demand, required energy sources, energy supply, and the energy sector in Germany until the year 2030.

Direct impacts affect the energy sector and encompass different steps, namely, generation, transport, storage, and distribution. These steps cover nanotechnological products and process innovations of energy technologies. Indirect impacts affect by contrast other sectors of industry via energy consumption through energy-saving products and applications. Figures 4.2 and 4.3 show examples for nanotechnological applications and technologies with direct and indirect impacts on the energy sector.

The approach to analyse the impacts of nanotechnological applications on the energy sector can be translated into five steps

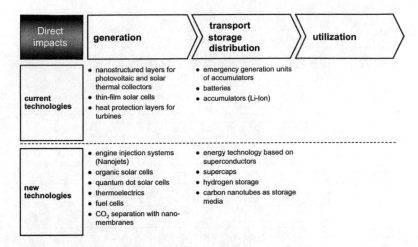

Figure 4.2   Direct impacts on the energy sector.

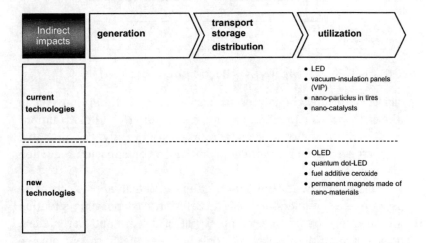

Figure 4.3   Indirect impacts of nanotechnology.

(see Figure 4.4). The first step is identification. It is to identify products, applications, services, and process innovation in all branches of industry and to estimate future fields of applications and substitutions.

Subsequent to the identification, influential nanotechnological applications considered to be of prime importance for the future are

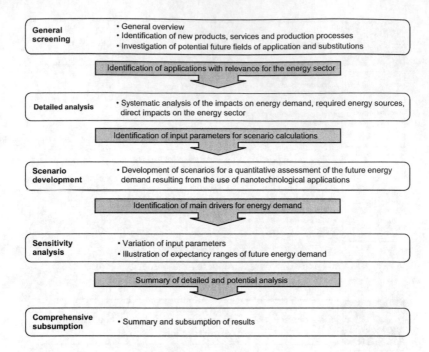

Figure 4.4 Methodical approach.

selected for further analyses, primarily in terms of their impacts on the energy sector (generation, transport, storage, and distribution) and on future energy consumption. Consequently, the potential impacts on efficient use of energy as well as on emission reduction are projected.

This is done with the help of scenario calculations. A scenario is a generally understandable description of a possible situation in the future based on a complex net of drivers and parameters. It can include the display of a development that directs from the present to a possible future situation (Gausemeier *et al.*, 1996). The projection is done with different parameters that are based on the approach of a multiple future to meet the uncertainties of the future development of nanotechnology. It opposes the development of energy demand with influence of nanotechnological applications to the present status quo. The development of the scenarios is differentiated in sectors and energy sources and is calculated in time intervals of five years until 2030.

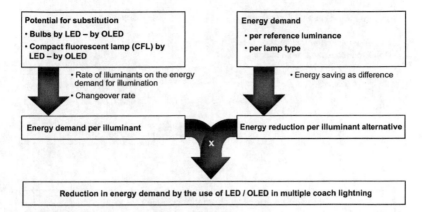

Figure 4.5 Calculation approach for the use of LEDs and OLEDs in multiple-coach lighting.

In order to explain the approach of the scenario analysis, two examples are used to describe the procedural method:

- The use of LEDs[1] and OLEDs[2] for multiple-coach lighting
- Insulation of buildings with innovative material (VIP[3])

Figure 4.5 illustrates the approach for the scenario LED and OLED. On the left-hand side, the potential for a substitution of conventional illuminants by LEDs and OLEDs is defined. The energy demand per illuminant for the substitution in due consideration of the rate of illuminants on the energy demand for illumination and the changeover rate is then derived. The possible energy reduction of LED and OLED for the generation of a reference luminance in comparison to conventional illuminants is calculated based on technical information for different illuminants.

The change in energy demand with the use of LEDs and OLEDs results from the multiplication of energy demand per illuminant and energy reduction per illuminant alternative.

Besides modification in heating technology, insulation of buildings can contribute to reduction in energy demand for space heating and therefore result in a great saving potential. Figure 4.6 shows

---

[1] LED: light-emitting diode
[2] OLED: organic light-emitting diode
[3] VIP: vacuum insulation panel

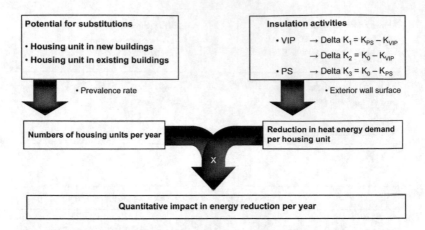

Figure 4.6 Calculation approach for innovative insulation materials.

the calculation approach. Beginning with the number of insulated housing units and with regard to the prevalence rate for VIPs, the numbers of housing units with VIPs can be calculated on a yearly basis. After that the heat transfer coefficient (k-value) of a sample wall is calculated based on technical parameters of insulation materials, for example, conventional material such as polystyrene (PS). The difference of the k-values is used to estimate the savings in space heating per square meter of exterior wall surface. On the basis of this and with the help of typical exterior wall surfaces of type-buildings, the energy savings for space heating for insulation with VIP compared to conventional insulation materials are calculated. The quantitative impact on the energy demand as a result is estimated by multiplying the numbers of housing units applying VIP with the resulting energy reduction.

This approach is adapted for other applications and products, and the scenarios are calculated for single applications and subsumed for expedient groups of products and applications.

The results of the scenario analyses oppose the development of energy demand through the influence of nanotechnological applications to the present status quo. Next to the scenario calculations (nano-scenario), a sensitivity analysis (nano−, nano+) with regard to the used input parameters is conducted in order to illustrate the expectancy range of future energy demand. Finally, the results of the general screening, the detailed analysis, the scenario calculations and the sensitivity analysis are summarised and evaluated. On the basis of this it is now possible to estimate the impact of selected nanotechnological applications on the energy sector in Germany and appraise the importance, possibilities, and potentials of nanotechnology as a whole for the energy sector, for changes in energy demand as well as for efficiency enhancements and emission reductions.

Bibliography

Gausemeier, J., Fink, A., and Schlake, O. (1996). Szenario-Management: Planen und Führen mit Szenarien, Carl Hanser Verlag, München, Wien.

4.2 Environmental Impact and Energy Demand of Nanotechnology

Michael Steinfeldt
Universität Bremen, Germany

Nanotechnology is frequently described as an enabling technology and fundamental innovation, i.e., it is expected to lead to numerous innovative developments in the most diverse fields of technology and areas of application in society and the marketplace. The technology, so it is believed, has the potential for far-reaching changes that will eventually affect all areas of life. Such changes will doubtlessly have strong repercussions for society and the environment and bring with them not only the desired and intended effects such as innovations in the form of improvements to products, processes and materials as well as economic growth in terms of new jobs for skilled workers, relief for the environment, and further steps towards sustainable business, but also unexpected and undesirable side effects and consequences.

In the following contribution, the focus is placed on the potential environmental relief provided by nanotechnology-based products and processes. Risk aspects, particularly in dealing with nanomaterials, are separately addressed in the Section 2.4.

4.2.1 *Environmental Reliefs Potentials of Nanotechnology*

Environmental relief potentials are understood here to include not only environmental engineering in the narrower sense (end-of-pipe technologies), but also and specifically process, production, and product-integrated environmental protection – thus not least the "input side" on the path to a sustainable economy: the reduction and modification of quantities (resource und energy efficiency) and properties (consistence) of the material and energy flows entering the techno sphere.

The analysis of new and existing nanotech products and processes in respect to environmental protection/pollution reveals a large and varied number of existing and potential areas of application (Figures 4.7 and 4.8); however, it must be noted that their environmental relevance has so far only been qualitatively represented. Quantitative investigations of anticipated or still to be realized environmental benefits arising from specific nanotechnological products and processes, as well as further-ranging environmental innovations such as product and production-integrated environmental protection or energy-related solutions have so far been the exception.

In addition to the potential applications in the area of end-of-pipe technologies, such as membranes (catalysis already extends

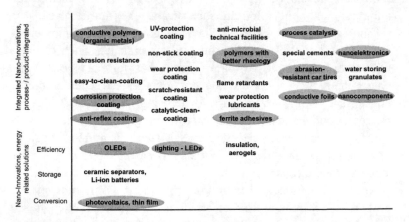

Figure 4.7 Nanotechnology-based products/applications on the market (Steinfeldt, 2010).

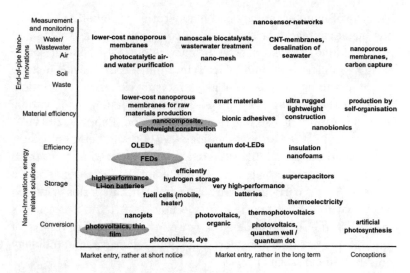

Figure 4.8 Anticipated nanotechnology-based applications (Steinfeldt, 2010).

beyond the purification of exhaust gases in many areas well into the area of integration), the following illustrations make clear that the most predominant and especially far-reaching potentials for nano-environmental innovation are to be found in the areas of integrated innovation and energy. In many fields of application numerous potentials for the realization of environmental benefits are opening up. Those applications for which separate case studies already exist and thus quantitative data about potential environmental benefits is available are indicated below with a light-grey background.

4.2.2 *Evaluation of Specific Application Contexts: Life Cycle Assessment*

As an assessment approach, the life cycle assessment (LCA) is the most extensively developed and standardized methodology for assessing environmental aspects and potential product-specific environmental impacts associated with the entire life cycle of a product. A workshop comprising international experts from the fields of both LCA and nanotechnology have concluded that the LCA ISO framework (ISO 14040:2006) is fully suitable to all stages of the life cycle of nanotechnology-based applications (Kloepffer, 2007). It has the advantage that by means of comparative assessments, an (extrapolative) analysis of eco-efficiency potentials in comparison to existing applications is possible. Typical impact categories include global warming/climate change, stratospheric ozone depletion, human toxicity, ecotoxicity, photo-oxidant formation, acidification, eutrophication, land use, and resource depletion (Rebitzer, 2004). At the same time, the LCA methodology – as with all methodologies – has its characteristic deficits: there are impact categories for which generally accepted impact models and quantifiable assessments do not exist. This is particularly true in the relevant categories of human and environmental toxicity. Furthermore, in LCA assessments the risks and the technological power (hazard) effectiveness of applications are not considered. A comprehensive methodology must provide for such analyses.

In the last years nanotechnology-based products and applications in LCA studies have been increasingly examined. A summary of studies of life cycle aspects identified is provided in Table 4.1. Currently, a large number of data gaps exist when considering the application of LCA to nanoproducts. Specifically, only minimal data detailing the material and energy inputs and environmental releases related to the manufacture, release, transport, and ultimate fate of nanocomponents and nanoproducts exists.

The accomplished studies have mainly focused on cradle-to-gate assessments. Cradle-to-gate is an assessment of a partial product life cycle from manufacture to the factory gate. The use phase and the after-use phase (recycling, disposal) of the product are usually omitted (Meyer, 2009).

For both phases almost no data regarding environmental impact exists, so that the environmental benefit could quite relate itself with a comprehensively LCA.

The results of the LCA comparisons make clear that nanotech applications neither intrinsically nor exclusively can be associated with the potential for a large degree of environmental relief. Nevertheless, for selected application contexts potentials for significant environmental relief can be ascertained using the chosen methods based on a comparative functionality of the different solutions.

Table 4.1   Overview of studies about life cycle aspects of nanotechnology-based applications

Nanoproduct	Approach	Tech benefits	Environmental benefits	Ref.
Anti-reflex glass for solar applications as compared with traditional glass	Not assessment, only indication of the environmental benefit	increased solar transmission	6% higher energy efficiency	(BINE, 2002)
Clay-polypropylene nanocomposite in light-duty vehicle body panels compared with steel and aluminium	Economic Input-Output Life Cycle Assessment (EIO-LCA)	reduced weight	overall reduced environmental impact; large energy savings	(Lloyd, 2003)
Nanoscale platinum-group metal (PGM) particles in automotive catalysts	Eco-profile following LCA methodology	reduced platinum-group metal (PGM) loading levels by 50%	overall reduced environmental impact (10–40%)	(Steinfeldt, 2003)
Photovoltaic, dye photovoltaic cells compared with multicrystalline silicon solar cells	Eco-profile following LCA methodology	dye photovoltaic cells with better energy payback time, but smaller efficiency		(Steinfeldt, 2003)
Carbon nanofibre (CNF) reinforced polymers	Ecobilan TEAM software	reduced weight, increased structural strength, improved conductivity	NA (not compared to traditional carbon fibers)	(Volz, 2004)
Nanoscale platinum-group metal (PGM) particles in automotive catalysts	EIO-LCA	reduced platinum-group metal (PGM) loading levels by 95%	overall reduced environmental impact	(Lloyd, 2005)
Carbon nanotubes	Substance flow analysis			(Lekas, 2005b)

(Contd.)

Table 4.1 (*Contd.*)

Nanoproduct	Approach	Tech benefits	Environmental benefits	Ref.
Ultradur® High Speed plastic as compared with conventional Ultradur	BASF Eco-Efficiency-Analysis	significantly improved flowability, reduction of working time and energy consumption of injection moulding process	reduced environmental impact (1.5–9%), only ozone depletion higher	(BASF, 2005); (Steinfeldt, 2010)
Desanilation, flow-through capacitor compared with reverse osmosis and distillation	Not assessment, only indication of the environmental benefit	increased energy efficiency	very high energy efficiency	(UBA, 2006)
Car tyre with nanoscaled SiO_2 and carbon black	Not assessment, only indication of the environmental benefit	increased road resistance	up to 10% lower fuel consumption	(UBA, 2006)
Nanocoatings compared with conventional coatings	Eco-profile following LCA methodology	necessary coating thickness smaller while maintaining functionality	5–8% higher resource efficiency 65% lower emissions of volatile organic compound	(Steinfeldt, 2007)
Styrene synthesis, CNT catalyst compared with iron oxide-based catalysts	Eco-profile following LCA methodology	change of type of reaction, reduction of the reaction temperature, change of reaction medium	ca. 50% reduced energy consumption of the synthesis process	(Steinfeldt, 2007)

Application	Method	Characteristic	Result	Reference
White LED and quantum dots compared with incandescent lamps and compact fluorescents	Eco-profile following LCA methodology	higher lifetime	compared with lamp higher energy efficiency, compared with fluorescent lamp only with light efficiency higher 65 lm/W	(Steinfeldt, 2007)
Organic LED displays and nanotube field emitter displays compared with CRT, liquid-crystal, and plasma screens	Eco-profile following LCA methodology	increased energy efficiency, higher display resolution, reduced display thickness	higher energy and resource efficiency; reduced material input by OLEDs; twice as high energy efficiency in the use phase	(Steinfeldt, 2007)
Ferrite adhesives compared with conventional adhesives	Eco-profile following LCA methodology	energy efficiently adhesives hardening by use of magnetic characteristics	12% (~40%) higher energy efficiency, dependent of size of adhered	(Wigger, 2007)
Polypropylene nanocomposite in packaging film, agricultural film, automotive panel compared with conventional films	Environmental and cost assessment	reduced weight, increased elasticity and strength of PP	for agricultural film lower impact for five out of seven environmental categories (35%); for packaging film and automotive panel very smaller or none benefit	(Roes, 2007)
Nano-delivery system compared with conventional micro-delivery system (vitamin E)	Screening LCA	higher penetration and conversion rates	potential for efficiency enhancement ca. 34%	(Novartis, 2007)
Car air filter with nanofibre coating compared with conventional car air filter	Environmental assessment, Umberto software	reduction of air resistance and the associated fan power	8% reduced energy consumption of fan, regarding the overall system environmental benefit very small	(Martens, 2008)

(Contd.)

Table 4.1 (*Contd.*)

Nanoproduct	Approach	Tech benefits	Environmental benefits	Ref.
Nanotechnology-based disposable packaging (nano-PET bottle) compared with conventional packaging	Environmental assessment, in particular CO_2 emissions	improved barrier characteristics in particular against oxygen	nano-PET bottle opposite to aluminium 1/3 and to glass of 60% fewer greenhouse gases	(Möller, 2009)
Manufacture of nanotechnology based solderable surface finishes on printed circuit boards compared with conventional surface finishes	Eco-profile following LCA methodology, Umberto software	necessary layer thickness smaller with same functionality	in relation to qualitatively comparable procedures depending upon environmental impact category around factors from 4 to 390 better	(Steinfeldt, 2010)
Nanotechnology-based (MWCNT) conductive foils as compared with conventional foils	Eco-profile following LCA methodology, Umberto software	necessary foils thickness smaller with same functionality	12.5–20% reduced environmental impact	(Steinfeldt, 2010)
Nanotechnology-based hybrid system city bus (Li-ion batteries) compared with conventional diesel city bus	Prospective Eco-profile following LCA methodology, Umberto software	reduction of fuel consumption by the hybrid system	ca. 20% reduced environmental impact by the future scenario	(Steinfeldt, 2010)

Source: Based on (Lekas, 2005a) and own data.

The reliability of the ascertained numbers is of course dependent on the quality and accessibility of the material and energy data available for the individual applications. For those nanotechnological processes still in development, only limited quantitative data is available. Likewise, when comparing established or mature technologies with those still in development, one must recognize that a new technology is at the beginning of its "learning curve" and holds the potential for significant increase in efficiency.

Due to the interdisciplinary nature of nanotechnology, an enormous wealth of methods for the production of nanoscale products can be found in the literature. Products can for example be differentiated according to their nanoscale basic structure: particle-like structures (e.g., nanocrystals, nanoparticles, and molecules), linear structures (e.g., nanotubes, nanowires, and nanotrenches), layer structures (nanolayers), and other structures such as nanopores. Materials can also be produced from the gas phase, from the liquid phase, or from solids in such a way that they are nanoscalar in at least one dimension.

The prevalence of diverse manufacturing routes for nanoproducts is a significant driver for nanotechnological innovations. All nanoproducts must proceed through various manufacturing stages to produce a material or device with nanoscale dimensions.

These techniques can be classified based on the type of approach in top-down or bottom-up. Top-down processes achieve nanoscale dimensions through carving or grinding methods (e.g., lithography, etching, and milling). Bottom-up methods assemble matter at the atomic scale through nucleation and/or growth from liquid, solid, or gas precursors by chemical reactions or physical processes (e.g., gas-phase deposition, flame-assisted deposition, sol-gel process, precipitation, and self-organization techniques).

4.2.3 *Evaluation of Specific Manufactured Nanoparticles*

The largest groups of manufactured nanoparticles for industrial applications are inorganic nanoparticles (e.g., TiO_2, ZnO, SiO_2, Ag), carbon-based nanomaterials (nanofibre, multi-walled carbon nanotubes [MWCNTs], single-walled carbon nanotubes [SWCNTs]), and quantum dots (semiconductor nanoparticles with a specific size, e.g., CdSe, CdS, GaN). Besides qualitative environmental assessments of the different manufacturing methods (Steinfeldt, 2007; Sengül, 2008), unfortunately, quantified material and energy flows, data exist also within this range only for a very small number of manufacturing processes and/or for individual nanomaterials. A summary of published studies is shown in Table 4.2. It is particularly remarkable that the majority of studies investigated the production of carbon-based nanomaterials.

Osterwalder and co-workers (2006) performed cradle-to-gate assessments of titanium dioxide (TiO_2) and zirconia dioxide (ZrO_2) nanoparticle production. The goal of the study was to compare energy requirements and greenhouse gas emissions for the classical milling process with that for a novel flame synthesis technique: flame-based nanoparticle production using organic precursors. The functional unit of the study was 1 kg of manufacturing materials.

Roes and co-workers (2007) evaluated the use of nanocomponents in packaging film, agricultural film, and automotive panels. The goal of the prospective assessment was to determine if the use of nanoclay additives in polymers (polypropylene, polyethylene, glass fibre-reinforced polypropylene) is environmentally more advantageous than conventional materials. Specific material and energy flows of the nanoclay production were collected. The manufacture of nanoclay includes several processes, e.g., raw clay (Ca-bentonite) extraction, separation, spray drying, organic modification, filtering, and heating.

Eckelman and co-workers (2008) performed an E-factor analysis of several nanomaterial syntheses, as the E-factor is a measure of environmental impact and sustainability that has been commonly employed by chemists. The E-factor (or waste-to-product ratio) includes all chemicals involved in production. Energy and water inputs are generally not included in E-factor calculations, nor are products of combustion, such as water vapour or carbon dioxide. Unfortunately, the results are not comparable with the other studies.

Kushnir and Sanden (2008) modelled the requirements of future production systems of carbon nanoparticles and used also a cradle-to-gate perspective, including all energy flows up to the production and purification of carbon nanoparticles. All calculations are made for a functional unit of 1 kg of nanoparticle. Several production

Table 4.2 Overview of studies of published LCAs of the manufacture of nanoparticles and nanocomponents

Nanoparticle and/or nanocomponent	Assessed impact(s)	Ref.
Metal nanoparticle production (TiO_2, ZrO_2)	Cradle-to-gate energy assessment, global warming potential	(Osterwalder, 2006)
Nanoclay production	Cradle-to-gate assessment, energy use, global warming potential, ozone layer depletion, abiotic depletion, photo chemical oxidant formation, acidification, eutrophication, cost	(Roes, 2007)
Several nanomaterial syntheses	E-factor analysis	(Eckelman, 2008)
Carbon nanoparticle production	Cradle-to-gate energy assessment	(Kushnir, 2008)
Carbon nanotube production	Cradle-to-gate assessment with SimaPro software, energy use, global warming potential,…	(Singh, 2008)
Single-walled carbon nanotube (SWCNT) production	Cradle-to-gate assessment with SimaPro software, energy use, global warming potential,…	(Healy, 2008)
Carbon nanofibre production	Energy use, global warming potential, ozone layer depletion, radiation, ecotoxicity, acidification, eutrophication, land use	(Khanna, 2008)
Nanoscale semiconductor	Cradle-to-gate assessment, energy use, global warming potential	(Krishnan, 2008)
Nanoscaled polyanilin production	Cradle-to-gate assessment with Umberto software, energy use, global warming potential,…	(Steinfeldt, 2010)
Multi-walled carbon nanotube (MWCNT) production	Cradle-to-gate assessment with Umberto software, energy use, global warming potential,…	(Steinfeldt, 2010)

Source: Based on (Meyer, 2009) and own data.

systems (fluidized bed chemical vapour deposition [CVD], floating catalyst CVD, HiPco, pyrolysis, electric arc, laser ablation, and solar furnace) are investigated and possible efficiency improvements are discussed. Carbon nanoparticles are found to be highly energy-intensive materials, on the order of 2 to 100 times more energy intensive than aluminium, given a thermal to electric conversion efficiency of 0.35.

Singh and co-workers (2008) performed environmental impact assessments for two potential continuous processes for the pro-duction of carbon nanotubes (CNTs). The high-pressure carbon monoxide disproportionation in a plug-flow reactor (CNT-PFR) and the cobaltmolybdenum fluidized bed catalytic reactor (CNT-FBR) were selected for the conceptual design. The CNT-PFR reactor has catalytic particles formed in situ by thermal decomposition of iron carbonyl. The CNT-FBR process employs the synergistic effect between the cobalt and molybdenum to give high electivity to carbon nanotubes from CO disproportionation.

Healy and co-workers (2008) investigated in the life cycle assessment three more established SWNT manufacturing processes: arc ablation (arc), CVD, and high-pressure carbon monoxide (HiPco). Each method consists of process steps that include catalyst prepara-tion, synthesis, purification, inspection, and packaging. In any case, the inspection and packaging steps contribute minimally to the overall environmental loads of the processes. Although the technical attributes of the SWNT products generated via each process may not always be fully comparable, the study provides a baseline for the environmental footprint of each process. All calculations are made with a functional unit of 1 g of SWNT.

Khanna and co-workers (2008) performed a cradle-to-gate assessment of carbon nanofibre (CNF) production. The goal of the assessment was to determine the non-renewable energy require-ments and environmental impacts associated with the production of 1 kg of CNFs. Life cycle energy requirements for CNFs from a range of feedstock materials are found to be 13 to 50 times higher than primary aluminium on an equal mass basis.

Krishnan and co-workers (2008) presented a cradle-to-gate assessment and a developed library of materials and energy require-ments and global warming potential of nanoscale semiconductor manufacturing. The goal of the study was to identify potential process improvements. The functional unit selected was 1 silicon wafer with a 300-mm diameter that can be used to produce 442 processor chips. The total energy required for the process is 14,100 MJ/wafer including 2,500 MJ/ wafer accounts for the manufacture of fabrication equipment. The greenhouse potential is 13 kg CO_2 eq/wafer.

Steinfeldt and co-workers (2010) performed several in-depth life cycle assessments of processes and products and also cradle-to-gate assessments of the production of nanoscaled polyaniline and of the production of MWCNTs. With the cooperation of firms, it was possible to produce detailed models of the manufacturing processes for nanoscaled polyaniline and for MWCNTs and to generate specific life cycle assessment data.

The data above can provide some insight regarding the potential burdens that must be addressed if the large-scale use of these types of nanoparticles and nanocomponents is to continue. For this purpose the data from the studies is expressed in a common mass-based unit. Accordingly, energy demand is presented in MJ equivalents/kg material and global warming potential is expressed as kg CO_2 equivalents/kg product. Energy consumption during the product life cycle is very important because it relates to the consumption of fossil fuels and the generation of greenhouse gases. Therefore, it is desirable to design manufacturing processes that minimize the use of energy. The data for energy consumption of the materials discussed above is shown in Figures 4.9 and 4.10. Additionally into the comparison data of conventional materials is included.

The represented cumulative energy requirements for various carbon nanoparticle manufacturing processes differ very strongly from each other. The various processes for the production of SWCNTs (excluding equipment fabrication) are by far the most energy-intensive processes compared with the production of other carbon nanoparticles. A cause for the very large differences between the examined studies lies in the different process conditions (temperature, pressure) of the manufacturing processes. Furthermore, large differences are found in the assumptions of reactions and purification yields. The relative small reference value appears remarkable for the mass production of carbon black by means of

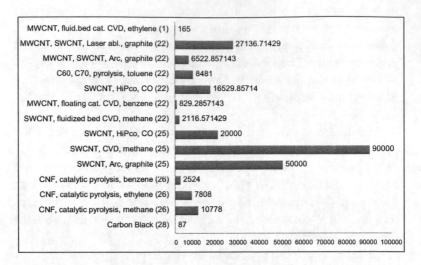

Figure 4.9 Comparison of the cumulative energy requirements for various carbon nanoparticle manufacturing processes [MJ equivalents/kg material]. See also Colour Insert.

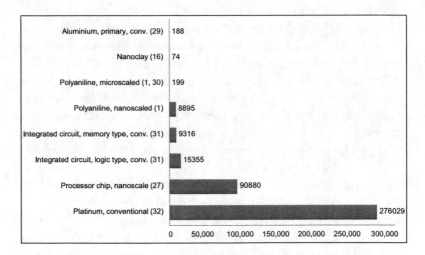

Figure 4.10 Comparison of the cumulative energy requirements for the production of various conventional and nanoscaled materials and components [MJ equivalents/kg product] (in parts own calculation). See also Colour Insert.

flame synthesis. The production of MWCNTs based on catalyst CVD surprises here with a relative small cumulative energy requirement.

The comparison of the cumulative energy requirements for the production of other conventional and nanoscaled materials and components make clear that the production of nanosemiconductors is also a very energy-intensive process. Only the extraction of the precious metal platinum as example is still more complex. The production of nanoscaled polyaniline is likewise very energy-intensive.

A comparison of the global warming potential for the production of various conventional and nanoscaled materials is shown in Figure 4.11. The CNF production has the largest impact when compared to the other materials. However, the production of nanoscaled polyaniline demonstrates also a high global warming potential. The reason for the larger global warming potentials for CNFs and polyaniline manufacturing is the much larger energy requirements when compared to other nanoparticle production. The production of MWCNTs based on fluidized bed catalyst CVD surprises here with a small global warming potential.

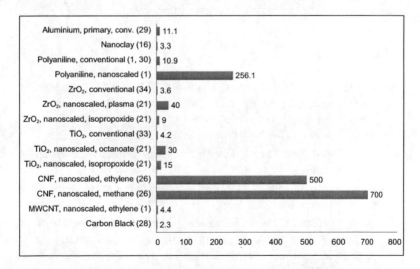

Figure 4.11 Comparison of the global warming potential for the production of various conventional and nanoscaled materials [CO_2 equivalents/kg product] (in parts own calculation). See also Colour Insert.

This represented results place no comprehensive life cycle assessments, and the results offer useful insight when considering the environmental impact of various nanomaterials and nanotechnology-based applications. It is commonly pointed out that the nanocomponent is only a fraction of the total product (often only 2%, 3%, or 4%) implying that only a small fraction of the environmental impact of a nanoproduct can be attributed to the nanocomponent and its manufacture. The high specific energy demand for the production of nanoparticle relates itself, then, in nanoproducts.

The results of the LCA comparisons make clear that nanotech applications neither intrinsically nor exclusively can be associated with the potential for a large degree of environmental relief. Nevertheless, for selected application contexts potentials for significant environmental relief can be ascertained using the chosen methods based on a comparative functionality of the different solutions (see Table 4.2).

Additional thought must also be given to address risk assessment and socioeconomic impacts/benefits which should be integrated with the LCA framework to provide a more comprehensive assessment tool for decision making when considering the use of nanoparticles for the manufacturing of nanoproducts.

Bibliography

Althaus, H. J. (2007). Aluminium, in Althaus, H. J., Blaser, S., Classen, M., Emmenegger, M. F., Jungbluth, N., Scharnhorst, W. and Tuchschmid, M. (eds), Life cycle Inventories of Metals, Ecoinvent Report No. 10, part I, Swiss Centre for Life Cycle Inventories, Dübendorf.

BASF AG (2005). Label Eco-Efficiency Analysis – Ultradur® High Speed. Ludwigshafen.

BINE Informationsdienst (2002). Antireflex glass for solar applications, Bonn (in German).

Chudacoff, M. (2007). Titanium dioxide, in Althaus, H. J., Chudacoff, M., Hellweg, S., Hischier, R., Jungbluth, N., Osses, M. and Primas, A. (eds), Life Cycle Inventories of Chemicals, Ecoinvent Report No. 8, Swiss Centre for Life Cycle Inventories, Dübendorf, pp. 761–768.

Classen, M. and Althaus, H.J. (2007). Platinum group metals (PGM), in Althaus, H. J., Blaser, S., Classen, M., Emmenegger, M. F., Jungbluth, N.,

Scharnhorst, W. and Tuchschmid, M. (eds), Life Cycle Inventories of Metals, Ecoinvent Report No. 10, part V, Swiss Centre for Life Cycle Inventories, Dübendorf.

Eckelman, M. J., Zimmerman, J. B., Paul T. and Anastas, P.T. (2008). Toward green nano: E-factor analysis of several nanomaterial syntheses, *J. Ind. Ecol.*, 12(3), 316–328.

Healy, M. L., Dahlben, L. J.and Isaacs, J. A. (2008). Environmental assessment of single-walled carbon nanotube processes, *J. Ind. Ecol.*, 12(3), 376–393.

Hischier, R. (2007). Carbon black, in Althaus, H.J., Chudacoff, M., Hellweg, S., Hischier, R., Jungbluth, N., Osses, M. and Primas, A. (eds), Life Cycle Inventories of Chemicals, Ecoinvent Report No. 8, Swiss Centre for Life Cycle Inventories, Dübendorf, pp. 173–178.

Hischier, R. (2007). Life Cycle Inventories of Packaging and Graphical Papers, Ecoinvent Report No. 11, part II, Swiss Centre for Life Cycle Inventories, Dübendorf, pp. 194–197.

Hischier, R. and Lehmann, M. (2007). Electronic components, in Hischier, R., Classen, M., Lehmann, M. and Scharnhorst, W. (eds), Life Cycle Inventories of Electric and Electronic Equipment: Production, Use and Disposal, Ecoinvent Report No. 18, part I, Swiss Centre for Life Cycle Inventories, Dübendorf, pp. 50–62.

Khanna, V., Bakshi, B. R. and Lee, J. (2008). Carbon nanofiber production: life cycle energy consumption and environmental impact, *J. Ind. Ecol.*, 12(3), 394–410.

Kloepffer, W., Curran, M. A., Frankl, P., Heijungs, R., Koehler, A. and Olsen, S. I. (2007). Nanotechnology and Life Cycle Assessment: Synthesis of Results Obtained at a Workshop in Washington, DC, 2–3 October 2006.

Krishnan, N., Boyd, S., Somani, A., Raoux, S., Clark, D. and Dornfeld, D. A. (2008). Hybrid life cycle inventory of nano-scale semiconductor manufacturing, *Environ. Sci. Technol.*, 42(8), 3069–3075.

Kushnir, D. and Sandén, B. A. (2008). Energy requirements of carbon nanoparticle production, *J. Ind. Ecol.*, 12(3), 360–375.

Lekas, D. (2005a). Analysis of Nanotechnology from an Industrial Ecology Perspective Part I: Inventory & Evaluation of Life Cycle Assessments of Nanotechnologies. Available at http://www.nanotechproject.org/15/analysis-of-nanotechnology-from-an-industrial-ecology-perspective (accessed 16 May 2007).

Lekas, D. (2005b). Analysis of Nanotechnology from an Industrial Ecology Perspective. Part II: Substance Flow Analysis Study of Carbon

Nanotubes. Available at http://www.nanotechproject.org/15/analysis-of-nanotechnology-from-an-industrial-ecology-perspective (accessed 27 June 2006).

Lloyd, S. and Lave, L. (2003): Life cycle economic and environmental implications of using nanocomposites in automobiles, *Environ. Sci. Technol.*, 37(15), 3458–66.

Lloyd, S., Lave, L. and Matthews, H. S. (2005). Life cycle benefits of using nanotechnology to stabilize platinum-group metal particles in automotive catalysts, *Environ. Sci. Technol.*, 39(5), 1384–1392.

Möller, M., Eberle, U., Hermann, A., Moch, K. and Stratmann, B. (2009). Nanotechnology in the Food Sector, Zürich (in German).

Martens, S., Eggers, B. and Evertz, T. (2008). Investigation of the use of nano-materials in environmental protection, Bericht zum Fachgespräch am 23.09.2008 in Dessau, Celle (in German).

Meyer, D. E., Curran, M. A. and Gonzalez, M. (2009). An examination of existing data for the industrial manufacture and use of nanocomponents and their role in the life cycle impact of nanoproducts, *Environ. Sci. Technol.*, 43(5), 1256–1263.

Novartis International AG, Ciba Spezialitätenchemie AG, Öko-Institut e.V., Österreichisches Ökologie Institut, Stiftung Risiko-Dialog (2007). CONANO - COmparative Challenge of NANOmaterials. A Stakeholder Dialogue Project, Projektbericht, Wien. Available at http://www.ecology.at/files/berichte/E11.565.pdf (accessed 11 January 2010) (in German).

Osterwalder, N., Capello, C., Hungerbühler, K. and Stark, W. J. (2006). Energy consumption during nanoparticle production: how economic is dry synthesis? *J. Nanopart. Res.*, 8(1), 1–9.

Primas, A. (2007). Zirconium oxide production from mineral sands, in Althaus, H. J., Chudacoff, M., Hellweg, S., Hischier, R., Jungbluth, N., Osses, M. and Primas, A. (eds), Life Cycle Inventories of Chemicals, Ecoinvent Report No. 8, Swiss Centre for Life Cycle Inventories, Dübendorf, pp. 861–875.

Rebitzer, G., Ekvall, T., Frischknecht, R., Hunkeler, D., Norris, G., Rydberg, T., Schmidt, W.-P., Suh, S., Weidema, B. P. and Pennington, D. W. (2004). Life cycle assessment: Part 1: Framework, goal and scope definition, inventory analysis, and applications, *Environ. Int.*, 30(5), 701–720.

Roes, A., Marsili, E., Nieuwlaar, E. and Patel, M. K. (2007). Environmental and cost assessment of a polypropylene nanocomposite, *J. Polym. Environ.*, 15(3), 212–226.

Sengül, H., Theis, T. L. and Ghosh, S. (2008). Toward sustainable nanoprod-ucts: an overview of nanomanufacturing methods, *J. Ind. Ecol.*, 12(3), 329–359.

Singh, A., Lou, H. H., Pike, R. W., Agboola, A., Li, X., Hopper, J. R. and Yaws, C. L. (2008). Environmental impact assessment for potential continuous processes for the production of carbon nanotubes, *Am. J. Environ. Sci.*, 4(5), 522–534.

Steinfeldt, M., Petschow, U. and Hirschl, B. (2003). Potential Applications of Nanotechnology based materials. Part 2: Analysis of ecological, social and legal aspects, Schriftenreihe des IÖW 169/03, Berlin (in German)

Steinfeldt, M., von Gleich, A., Petschow, U. and Haum, R. (2007). Nanotech-nologies, Hazards and Resource Efficiency. Springer, Heidelberg.

Steinfeldt, M., von Gleich, A., Petschow, U., Pade, C. and Sprenger, R. U. (2010). Environmental Relief Effects through Nanotechnological Processes and Products, UBA-Texte 33/2010, Dessau (in German).

Umweltbundesamt (UBA) (2006). Nanotechnology: Prospects and Risks for Humans and Environment, Berlin. Available at http://umweltbundesamt.de/uba-info-presse/hintergrund/nanotechnik.pdf (accessed 13 January 2007) (in German).

Volz, S. and Olson, W. (2004). Life cycle assessment and evaluation of environmental impact of carbon nanofiber reinforced polymers. Submitted to the Journal of Industrial Ecology on August 2, 2004.

Wigger, H. (2007). Nanotechnological and bionic approaches in the application field adhesive bonding and their potential environmental relief effects, Studienarbeit an der Universität Bremen, Bremen (in German).

4.3 Potentials of Nanotechnology for Improvements in Energy Efficiency and Emission Reduction

Jochen Lambauer, Dr. Ulrich Fahl, and
Prof. Dr. Alfred Voß
Institut für Energiewirtschaft
und Rationelle Energieanwendung (IER),
Universität Stuttgart, Germany

In order to evaluate possible impacts of nanotechnological applica-tions on the energy sector in Germany, different applications and products are analysed in detail. Based on this information, scenario calculations are developed. With their help the development of energy generation and energy consumption is projected till 2030 for Germany. Scenario and sensitivity analyses are calculated and the analysed products and applications are described in the following based on the value chain of the energy sector.

4.3.1 *Energy Sources and Conversion*

In the field of energy sources, possible impacts on energy generation based on renewable energies are evaluated. In particular, the possible impacts and potentials of nanotechnology on the use of solar energy are matter of the analysis.

4.3.1.1 Solar heat and photovoltaics

In case of solar energy different systems for utilisation are possible. It is divided into different systems to utilise solar heat, so-called solar thermal systems, where sunlight is used as heat source for heating and cooling purposes, and photovoltaics where sunlight is directly converted into electricity.

Solar heat can be used for tap water heating, space heating in buildings, or on an industrial scale for power generation in solar power plants. Thereby sunlight is concentrated, for example, by parabolic mirrors, and used to generate steam, which spins a turbine for electricity production.

Nanotechnology shows great potential in the field of solar energy. Next to the further development of existing conventional solar cells based on silicon, there are a wide variety of new promising solar technologies (Joop, 2003). For example, dye solar cells, polymer solar cells, quantum dot solar cells, organic solar cells, and thin-film solar cells count among these new technologies.

Nanotechnology can enable not only the necessary break-throughs, but also efficiency enhancements in conventional technologies (Paschen *et al.*, 2004), even though it is primarily about functionalisation with the help of nanostructured coatings. Next to an enhancement in efficiency it is expected that production costs of photovoltaics can be reduced considerably.

4.3.1.1.1 Antireflective coatings In the application of solar cells, the loss of solar radiation amounts to around 8% by reflection and a further of 2% by absorption and dispersion. Taking into account all different possible losses, the solar transmission is merely 90% of solar radiation (Hofmann, 2006). With the help of nanoscale functional coatings that are antireflective and therefore increase transmission of glass the efficiency of solar panels and photovoltaics heightens (Hartmann, 2006). Antireflective coatings consist of porous silicon dioxide (SiO_2). SiO_2 particles (with diameters of 20 to 50 nm) are applied on glass substrate with sol-gel proceedings and afterwards thermally hardened at temperatures of around 650°C. At optimal layer thickness of 150 nm surface reflection of glass can be reduced from 8% to 2% so that nearly the total spectrum of solar radiation (from 400 nm to 2500 nm) can be energetically used (HA, 2005). Antireflective coatings can be utilised for photovoltaic and solar panels (heating and tap water) as well as for high-temperature receivers for concentrating solar power (CSP) plants (Kuckelkorn, 2006; Uhlig, 2007).

Efficiency of photovoltaic panels can be increased by 3.5 percentage points up to 4 percentage points and for solar panels about 5 to 6 percentage points (Hofmann, 2006). In CSP plants the absorption tube that should show high light transmission in order to minimise reflection losses has to withstand extreme conditions (fluctuations in temperature of around 400°C). In order to do so, borosilicate glass is used in most applications. One main disadvantage is that on each boundary layer around 4% of the sunlight is reflected. Altogether around 8% of sunlight is lost due to reflection. Conventional antireflective coatings could not be used on borosilicate glasses (Helsch, 2006).

A research team at the University of Clausthal was able to develop a wear-resistant nanoscale antireflective coating, which leads to an increase in light yield of about 5 percentage points. Light transmission could be optimised from 92% up to 97% (Helsch, 2006; IHK-Braunschweig, 2006).

Other advantages through the use of TiO_2 particles are an increased surface hardness as well as scratch and wear resistance (Hartmann, 2006). Further possibilities to increase efficiency for the use of solar energy are so-called prismatic selective layers. With the help of these layers, only the usable part of the solar spectrum could be coupled into the solar cells in order to avoid the heating up (Paschen *et al.*, 2004). Coatings with silver nanoparticles, for example, could provide the needed selectivity (Grüne *et al.*, 2005). Simulations for the analysis of radiation absorption show that, for example, prismatic selective layers based on silver nanoparticle could result in a potential increase of efficiency of photovoltaics by around 24% (Paddon and Bernhard, 2008).

4.3.1.1.2 New photovoltaic technologies Nanotechnology offers several new possibilities in the field of photovoltaics. Some possible innovations are described hereinafter.

Thin-film solar cells

Thin-film solar cells consist of thin semiconducting layers with thicknesses of 1 μm and therefore around factor 100 thinner than mono- or multicrytalline silicon solar cells. Mainly they are manufactured based on amorphous silicon (a-Si), cadmium telluride (CdTe), and copper-indium-(gallium)-diselenide (Cis or CIGS). Thin-film solar cells show the advantage that they require, based on their thickness, considerably less material for manufacturing and they show still similar efficiencies. Other advantages are that the efficiency stays almost the same in suboptimal lighting conditions (cloudiness, incidence angle of radiation) and temperature variations. In comparison to silicon wafer technology, thin-film solar cells allow low-temperature manufacturing processes, integrated cell wiring, and high level of automation in batch production. Therefore manufacturing costs decrease and new possible fields of application, such as facades, flat roof, or the integration in textiles, arise (Luther, 2008; Quaschning, 2008).

According to Quaschning (2008) and Reisch (2007), a-Si thin-film solar cells reach efficiencies of around 6% to 8%, but long-term stability is still a problem. In comparison to crystalline solar cells, they cover the possibility to be used on facades and thereby show almost no decrease in efficiency. Transparent a-Si thin-film solar cells could be used, for example, as roof facades in order to allow natural ambient light (Hagemann, 2007). Cadmium telluride thin-film solar cells show high efficiencies of around 16.5% on laboratory scale and could reach even up to 19% in the near future (Lüdemann *et al.*, 2005). On the other hand, the toxicity of cadmium appears to be a barrier (Quaschning, 2008). Right now Cis/CIGS thin-film solar cells show the highest efficiencies of up to 11.8%. On laboratory scale efficiencies of almost 20% could be reached (Quaschning, 2008).

Quantum dot solar cells

Quantum dot structures that could enable an ideal utilisation of solar radiation within the infrared spectrum, too, are in the state of fundamental research (Paschen *et al.*, 2004). In the case of quantum dots thermal relaxation of stimulated charge carriers as well as electronic conditions change. There exist different concepts in order to utilise these properties within quantum dot solar cells. According to Texocon (2008) the potential for the energy sector is regarded as promising despite the different technological difficulties in the realisation of such kind of solar cells. With the help of quantum dots on the one hand it is possible to create several electron hole pairs per proton and on the other hand absorption bands can be ideally adapted to the wavelength of the irradiated light (Luther, 2008). The use of quantum dots shows great promise for the use in thin-film and staple solar cells. Theoretically the efficiency of conversation could rise up to 60%. Already in the next 10 years quantum dot solar cells could be ready to enter the market, but at the beginning, relative low efficiencies of around 17% are expected. Staple solar cells could reach efficiencies of around 30% in the coming 20 years (Green, 2008).

Dye solar cells

The physical principal of dye solar cells or Grätzel cells, which are named after the inventor Prof. Michael Grätzel, were developed in the beginning of the nienties and are based on the principal of photosynthesis. Dye solar cells with solid electrolyte materials (so-called Polymer TiO_2 solar cells) reach efficiencies of around 4% (Oey *et al.*, 2006). According to BSI (2007) nanocrystalline coated cells can convert up to 12% of solar energy into electrical energy. At the moment, efficiencies are between 5% and 8% (BSI, 2007; Tolbert, 2008). Dye solar cells show high mechanical flexibility and can be manufactured partly or fully transparent. Moreover, it is possible to colour the cells in order to use them as active windows that produce electricity, too (Wevers *et al.*, 2002). One main advantage of dye solar cell is that the effciency stays almost the same in low lighting conditions. This feature allows efficiencies of up to 13% at cloudiness, arenaceous air or at indoor use (Luther, 2008).

Organic solar cells

Organic solar cells show mechanical flexibility paired with low manufacturing costs (FhG-ISE, 2006). They can be divided into two different types of cell buildup. Within the bi-layer concept the photocatalytic layer consists of a plane *p-n*-boundary layer made of organic *p-* and *n*-conduction materials. In the composite concept the photocatalytic layer consists of a nanophasic mixture of these materials. In both concepts the thickness of these photocatalytic layers is around 100 to 200 nm (Grüne *et al.*, 2005). Based on WFU (2007) organic solar cells already reached efficiencies of more than 6%. Researchers hope that efficiencies could reach up to 10% in the coming years. Main research is conducted in the field of efficiency increase, manufacturing costs reduction, and the further development of long-term stability of organic solar cells (ISE, 2007).

Staple solar cells

In comparison to other photovoltaic cells staple solar cells show the highest theoretical conversion efficiencies in laboratory tests.

They consist of a combination of several functional layers. The advantage of the combination of various semiconducting layers is the high yield potential based on the different photonic band gaps.

Therefore it is possible to convert a greater light spectrum into electricity. On laboratory scale these semiconducting systems reach efficiencies of up to 40%. Due to high manufacturing costs, stable solar cells are just used in aerospace applications for the time being (Brand *et al.*, 2007; ISE, 2009; Luther, 2008; ObservatoryNano, 2008).

4.3.1.2 Fuel cells

In principle all fuel cells consist of two porous layers, the catalytic active electrodes (anode and cathode) that are separated by a third layer the electrolyte. There are different types of fuel cells that differ in assembly, used electrolyte, and operating temperature (Grüne *et al.*, 2005). Fuel cells can be divided in low, medium, and high temperature fuel cells. Low temperature fuel cells are, for example, alkaline fuel cells (AFC) and polymer electrolyte fuel cells (PEFC, PEM) that operate in a temperature range of 60°C to around 120°C. Phosphoric acid fuel cells (PAFC) operate from 160°C to 200°C. High-temperature fuel cells (500°C to 1000°C) are, for example, molten carbonate fuel cells (MCFC) and solid oxide fuel cells (SOFC). For more information please see, for example, Aigle and Jörissen (2003), Blum and Peters (2002), or Gummert and Suttor (2006). Diffusion of fuel cells is limited mainly by high costs. But improvements of membranes, catalysts and electrodes based on nanotechnology can make fuel cells more powerful and cheaper (Paschen *et al.*, 2004).

4.3.1.3 Fuel additives

On account of high surface area of nanoparticles, they are qualified as catalysts to enhance efficiency and effectiveness of chemical reactions. In combustion processes, nanoporous catalysts or nanoparticles can enhance efficiencies and reduce emissions. As stated by Nanoforum (2004), energy saving potentials are enormous. 'It has been recognised for some time that cerium oxide could give a cleaner-burning fuel, but until nanoparticles could be manufactured, the catalyst simply settled out on the bottom of the gas tank' (Edwards, 2006). The effectiveness of cerium oxide is influenced by the transformation from CeO_2 to Ce_2O_3 and the ability to release oxygen by combustion. Through oxygen the fuel volume can be enhanced and in addition burning temperature is lowered. As a result existent soot accumulations in the engine can be burned off (Fairley, 2008; Oxonica, 2005a; Oxonica, 2005b; Oxonica, 2007, 2008).

4.3.1.4 Nanostructured membranes

Fields of application for nanomembranes are, for example, purification and conditioning of water or desalination of seawater. Hereby membrane layers with nanosize pores are able to exfiltrate bacteria using mainly ceramic materials. Different choices of membrane material, for example, polymers, metal alloys, or carbon nanotubes, can cater for other possible applications.

Nanomembranes can be applied to fuel cells or to osmotic power plants as well as to carbon capture in the field of power plant engineering (Jopp, 2006). In comparison to chemical deposition techniques, membrane technologies show lower losses in efficiency. According to (Birnbaum *et al.*, 2010) it is assumed that net efficiency of actual coal power plants is lowered by about 8 to 10 percentage points through the use of carbon capture and storage (CCS). According to Viebahn *et al.* (2010) losses in net efficiency of 4 percentage points are conceivable in future.

As a result energy consumption of a power plant increases by 12% to 30%. It is believed that further development of membrane technologies can lower losses in efficiency. The increase of own energy demand of power plants shall be limited to 8% (ETAP, 2007). For the CCS technology oxyfuel combustion losses in efficiency of around 10% in comparison to actual power plants without CCS are expected. By realisation of air separation with ceramic high temperature membranes losses in efficiency could be reduced to about 2% to 5% (Birnbaum *et al.*, 2010; Radgen *et al.*, 2006; RWTH, 2008; Viebahn *et al.*, 2010).

4.3.1.5 Thermoelectric generators

By means of thermoelectric generators (TEGs), temperature differences can be converted directly into electric energy by utilising the so-called Seebeck effect. Generally TEGs consist of semiconducting materials. With the help of new materials, for example, nanocrystalline layers made of semiconducting material, high efficiencies and distinctive thermoelectric conversion becomes possible (BMBF, 2004; Paschen *et al.*, 2004).

Efficiency of conversion depends mainly on the thermoelectric effectiveness of the materials used. At present, the use of solid materials with best possible thermoelectric effectiveness result in z-values of 1 (Luther, 2008) and efficiency of such generators ranges from 5% to 10% (Fröhlich, 2007).

Indeed enhancements in efficiency of around 50% to 200% are expected with the help of nanotechnology (UH, 2008). TEGs could be used to generate electricity by waste heat at power plants or in industrial processes (Grüne *et al.*, 2005). Based on the findings of Pester and Trechow (2009) TEGs could be used in automotive industry already in 2012 to convert exhaust heat into electricity.

4.3.2 *Energy Storage and Distribution*

In the field of energy storage and distribution, the use of supercapacitors is evaluated. Nanoporous electrode materials can be used in supercapacitors (supercaps) that are applied as interim-storage devices in different applications. If a supercapacitor is charged, energy in the form of concentration of electrons or charge is stored in an electrochemical double layer on the surface of the material. Energy density of supercapacitors is up to hundredfold higher than that in conventional condensators and power density is tenfold higher than that in accumulators. It shows that they have proven reliability and are maintenance-free (Fischle, 2005; Grüne *et al.*, 2005; Nanoforum, 2004). In order to gain a distinct increase in capacity in comparison to conventional capacitors, the electrodes have to show a very huge surface. Moreover, a high charge storage capability for electrochemical interface is required while for electrodes carbon with pores in nanoscale is used (Nanoforum, 2004).

Industrial soot has so far been used as electrode material. Alternatively, carbon aerogels, which show a very huge inner surface as well as controllable pore distribution and pore diameters, can as well be used.

The use of multi-shelled carbon nanotubes as well as fullerenes is matter of research. Increase in surface area fostered by the nanoporous morphology should result in a higher accumulation of electric charge in the electrochemical double layer and in future energy densities of above 10 Wh/kg are probable (Grüne *et al.*, 2005; VDI-TZ, 2006). According to Bullis (2011) a US company has developed supercapacitors with an energy density of 20 Wh/kg. The aim are values from around 35 to 40 Wh/kg.

Supercapacitors can be used for voltage support to cover capacity peaks or to preserve and support accumulators. Promising is the combination of supercapacitors, with high performance and high capacity density, with accumulators that show great energy densities. Possible applications are hybrid vehicles, interruption-free power supply, support of subway grids, energy cache in photovoltaic applications, or slip regulation in wind power plants (Fischle, 2005; Grüne *et al.*, 2005).

4.3.3 *Energy Use*

Up to now nanotechnological applications are already on the border between research and introduction on the market, as well as being counted to the field of energy use and therefore have indirect impacts on the energy sector. A number of different applications and their impact on energy use are the focal points of the discussion. It covers, for example, from new developments in illumination, display technology, to nanoparticles in chemical industry and tyre production or new insulation materials.

4.3.3.1 LED and OLED in illumination

LEDs are semiconductor light sources that produce light in nanometre-thick semiconducting layers, where an *n*-conductive area (negative charge), a *p*-conductive area (positive charge) and a pn change-over (depletion layer) are situated.

By impressing voltage electrons in the *n*-area, the depletion layer can be avoided and positive charges in the *p*-area can be connected. In this process, energy is emitted in form of electromagnetic radiation and is released. Depending on the semiconducting material, light in a specific colour (red, green, yellow and blue) is emitted

(FGL, 2006; LED-Info 2006). The application of LEDs has been increasingly discussed recently as an alternative to conventional light sources, since it is believed that LEDs can be used more efficiently for light generation. The challenge of LEDs is that just around 20% to 30% of the produced light can be uncoupled of the component although they convert up to 90% of energy into light. At the moment luminance of LEDs can be compared with conventional energy saving bulbs or halogen bulbs.

In order to achieve a wider market penetration in multiple-coach lighting, a light efficiency of 200 lm/W is needed and shall be realized till 2020. Further possible applications are automotive lighting and street lighting (BG-ETEM, 2011; BMBF, 2011; Edwards, 2006; FGL, 2006; Jüstel, 2006; OIDA, 2002; Steinfeldt *et al.*, 2004).

Organic light emitting diodes (OLEDs) are components based on organic layers that emit light by impressing a voltage. The layers can be superimposed on top of each other on flexible and transparent material (e.g., polyethylene) or on a rigid carrier substrate such as glass. The actual light emitting layers consist of organic material with a thickness of a few nanometres and the selection of the material defines the fluorescent color. OLEDs have advantage over light emitting diodes made of inorganic semiconducting material since OLEDs can be manufactured in almost every colour based on the chemical variability; in other words the thin-film systems can be applied on areal and flexible carriers and the components can be manufactured potentially cheap. Moreover, they are very energy efficient in comparison to conventional light bulbs. Attributed to their advantageous properties, OLEDs can be used as displays and light source. Although the first applications have been available on the market, there are still rooms for improvement. In fact, OLEDs can be destroyed by contact with water and oxygen, and the polymer that produces blue light loses its luminosity already after a short time (BMBF, 2006; Hartmann, 2006; Paschen *et al.*, 2004).

According to OLLA (2009) solid-state lighting technologies (LED and OLED) can be seen as an efficient possibility to convert energy directly into visible light. LEDs are the best solution for direct lighting, whereas OLEDs are promising technology for large-scale lighting. OLEDs are the first thin and planar light source that generate a uniform illumination without high heat generation.

OLEDs have a lifetime of more than 10,000 hours and achieved a light yield efficiency of 50 lm/W in research projects. However, in order to survive in the mass market, light yields of 100 lm/W and a lifetime of 100,000 hours are necessary (OLLA, 2009). A research project is taking place with the aim of reaching these values (OLED100 2009, 2011).

4.3.3.2 New display technologies

The market of display technologies is dynamic and prosperous. According to Yersin (2010), a drastic change in display and illumination technologies emerges. Several years ago CRTs were dominating the display technologies market. Since a few years the rate of liquid-crystal displays (LCDs) has been growing steadily. According to Stobbe *et al.* (2009), LCD and CRT displays had in each case almost the same market share of around 14 million in the household sector in 2007. However, CRT displays play gradually a less important role.

New display technologies based on nanotechnology are, for example, field emission displays (CNT-FEDs) based on carbon nanotubes and OLED displays. With OLED displays it will be possible in the future to manufacture planar displays or light sources with a thickness less than 0.5 mm (Yersin, 2010). This technology could possibly enable wall papers to be functioned as light sources and could contribute to the field of display technologies in a way that colour displays could offer properties which have not yet been achieved – e.g., colour fastness, brightness, angle of view, weight, and energy consumption. It is assumed that simple and low-cost manufacturing processes such as silk screen printing or inkjet printing can be used. OLED displays show a simple structure with one or several organic films (typically with a thickness of 100 to 200 nm) between two electrodes.

Usually two types of organic material are used: long-chain polymeres that are processed out of a solution and small molecules that are thermically evaporated under vacuum. At a current of 3 to 5 V, electrons are injected in the film with a cathode material that shows low electron affinity. Possible cathode materials are metals such as barium, calcium, or specific fluorides. At the same time holes (positive charge) are injected in the organic materials with the help of a transparent anode (e.g., indium tin oxide [ITO]) with a high electron affinity. By impressing an electrical field holes and electrons migrate through the organic film and create under recombination excited states that break down and emit photons and the light is decoupled through the transparent ITO electrode. A determining factor for the efficiency of OLEDs is the used illumination material (emitter molecules). The possible use of existing manufacturing technologies could lead to a fast and cost-efficient market entry (Yersin, 2010).

CNT-FEDs are still in a state of research. The main components of a field emission display are the rearward panel with a cathode layer, a vacuum gap, and a front panel that is coated on the inner surface with an ITO layer as anode. One of the most important components of a CNT-FED is the cathode layer (emitter layer) on the rearward panel because the characteristic, nature, and constitution of the emitter are decisive for picture quality, energy consumption, and lifetime. Based on the requirements of FEDs such as low diameter, stability at high current densities, high conductivity, low energy losses, and great chemical stability, carbon nanotubes are exceedingly appropriate for the application in FEDs. FEDs in general function similar to CRT displays with several parallel emitting cold field electron sources per picture element and therefore achieve similar characteristics such as conventional CRT in regard to brightness, angle of view, or colour reproduction. A great advantage of FEDs is that by impressing an electrical field, a cold electron discharge is generated with much less energy compared to conventional CRTs that have to be heated up (Steinfeldt *et al.*, 2004, 2007).

The advantages of OLED and CNT-FED displays and the resulting ecological effects according to Steinfeldt *et al.* (2004, 2007) are summarised in Table 4.3.

Steinfeldt *et al.* (2004, 2007) compare different display technologies in terms of energy consumption during usage phase. It becomes obvious that OLED and CNT-FED displays show great advantages in regard to energy consumption (see Table 4.4).

Table 4.3   Advantages of OLED displays and CNT-FED

Advantage	Ecological effect
OLED	
Simple configuration, reduction of manufacturing steps	Decrease of complexity of manufacturing facilities
Lower display depth (1.8 mm compared to 5.5 mm) less material inventory	Increase of material efficiency
Lower evaporation temperature of organic molecules (300°C to 500°C) compared to metals (1 200°C)	Decrease of energy consumption during manufacturing
Organic luminescent substances	Problem with long-term stability
Lower specific energy consumption during usage phase	Increase of energy efficiency
CNT-FED	
Lower specific energy consumption during usage phase	Increase of energy efficiency
Lower display depth (3.5 mm) less material inventory	Increase of material efficiency
Very high resolution	
Use of carbon nanotubes (CNT)	Complex manufacturing process (selective partial growth of CNT)

Table 4.4   Development of energy consumption of different display technologies

Year Dimension	Technology	2000 32 in.	2005 50 in.	2010 50 in.
Energy consumption	LCD	140 W	120 W	100 W
	PDP[4]	300 W	200 W	100 W
	CNT-FED	70 W	70 W	40 W
	OLED	–	60 W	30 W
	CRT	200 W	230 W (HDTV[5])	200 W (HDTV)

[4] PDH: plasma display panel
[5] HDTV: high-definition television

Furthermore, energy consumed along the whole life cycle cannot be overlooked. Energy consumption of display technologies in each stage of lifetime and the total energy consumption as stated by Steinfeldt *et al.* (2004, 2007) are shown in Table 4.5. Two different variations of OLED displays are examined. Resulting from the simple configuration of OLED displays in preproduction and

Table 4.5   Energy consumption of display technologies

Technology	Preproduction (MJ)	Manufacturing (MJ)	Usage (MJ)	Total (MJ)
CRT	366	18 300	2 290	20 956
LCD	633	1 440	853	2 926
PDP	633	1 440	1 422	3 495
OLED 10%	570	1 296	427	2 293
OLED 30%	443	1 008	427	1 878
CNT-FED	570	1 440	498	2 508

manufacturing, lower energy consumptions can be assumed (10% and 30% decrease in energy consumption).

It becomes obvious that OLEDs show the lowest total energy consumption based on the low energy consumption during usage. Also for the other stages of life cycle, OLED technologies show advantages relating to energy. According to Steinfeldt *et al.* (2004, 2007) the assumed reductions in energy consumption of OLEDs in preproduction and manufacturing are possible but in order to succeed it is also necessary to overcome problems with long-term stability of the luminescent substances. CNT-FEDs have a lower energy consumption than conventional technologies such as LCDs, too, when the assumed energy consumptions in preproduction and manufacturing can be achieved. It can be concluded that nanotechnologies have the potentials for energy saving in display technologies.

By means of their extraordinary optical properties, quantum dots can radiate light by stimulation and therefore represent a new display technology. They produce a very pure and bright light and are still able to cover a whole range of colours. In comparison to LCDs they need just one-thirtieth of the electricity consumption. This fact is attributable to the abdication of the background lighting. Up to now quantum dot displays have just existed on the laboratory scale; nevertheless, it can be assumed that quantum dot technologies take their part in the display technology market in the long term (Bullis and Wirth, 2006; Steinfeld *et al.*, 2004). According to Patel (2011) a first full-colour display with quantum dots was manufactured by researchers from Samsung Electronics.

4.3.3.3 Ultra-high-performance concrete

Considering the importance and size of the construction sector, energetic changes in the use of the basic material concrete can result in great impacts on the energy demand. In regard to the energy demand, it should be noted that most energy is used for the manufacturing of the binder cement and not for the building material concrete. Cement production, mining of rocks, crushing, and calcination of cement use the most energy (BDZ, 2010).

Ultra-high-performance concrete (UHPC) is a concrete class that shows compressive strength above strength category C 100/115 as defined in DIN EN 206-1. At the University Kassel a high-performance concrete was developed that allows, based on its steel-like compressive strength (200 N/mm^2), delicate, light, sustainable, and yet highly stable and corrosion-resistant constructions. The high performance is basically attributable to the addition of easy solvent and highly reactive nanoscale silicic acid (SiO_2). Furthermore, particle composition of the concrete is optimised on micro- and nanoscale by aimed addition of further mineral fillers so that the structure of the solidified cement stone becomes extraordinarily dense and solid (Luther, 2007a). Thereby compressive strength and other important properties for construction are improved. Table 4.6 compares selected properties of conventional concrete and UHPC.

It is assumed that when using UHPC for the same building element, about 60% of raw materials and about 40% of energy can be saved. It has to be considered that for 2009, the German cement manufacturing alone resulted in an energy demand of around 100 million GJ (BDZ, 2010; BDZ, 2011; DFG, 2004; VDZ, 2007).

In regard to today's costs of UHPC of around 700 to 1000 €/m^3 compared with conventional concrete (50 to 75 €/m^3), possible pilot applications are in specialised building construction and in bridge engineering (Luther, 2007a).

Table 4.6   Selected properties of conventional concrete and UHPC

Property	Unit	UHPC	Conventional concrete
Density	(kg/dm^3)	2.45–2.55	2.2–2.5
Compressive strength	(MPa)	180–200	10–60
Bending strength	(MPa)	36–40	2–8
Tensile strength	(MPa)	8–10	1–4
Modulus of elasticity	(GPa)	55–60	20–40

4.3.3.4 Insulation with vacuum-insulation panels

Vacuum-insulation panels (VIPs) are nanoporous silicic acid boards that are wrapped in adequate foil systems.

During manufacturing these highly dispersive, pyrogenous silicic acid agglomerates, a nanoporous and open structure develops that minimises mobility of the enclosed air molecules and hence the transmission of heat energy via convection. In addition silicic acid has the beneficial property of bonding foreign atoms, for example, water (Randel, 2003). The very low heat conductivity of VIPs at standard temperature and standard pressure is lowered by factor 4 with the help of evacuation. It is possible to get a heat conductivity of 0.0053 W/mK.

This means that with VIPs it is possible to get values of heat conductivity that are about factor 5 to 10 lower than those for conventional insulation materials. Or vice versa, the same insulation effect can be achieved with an insulation material that is fivefold to tenfold thinner (BINE, 2001, 2004, 2011; BFE, 2003; va-Q-tec, 2009).

With the help of new manufacturing methods and innovative foil technologies the operational reliability of vacuum technology can be guaranteed, and it becomes of interest to the building sector, too. But vacuum technologies for insulation applications in the building sector have to meet specific conditions, for example, a lifetime of 50 years. Furthermore, for the handling of VIP there are more restrictions than for conventional insulation materials, for example, the covering of the panels must not be damaged (vip-bau, 2008).

In 2007 after several test applications VIPs were generally approved by the building authorities. The heat conductivity value is defined as 0.008 W/mK, including deterioration effects and thermal bridge effects on the edging of the panels (va-Q-tec, 2009).

Other applications than insulation of buildings are in logistics (insulation of containers, tanks) or for the exploitation of so far unexploited energy sources. At deep-sea oil exploitation low temperatures lower viscosity of the oil, so that conveying is hindered. With the use of double-walled pipes that are insulated with nanoporous silicic acid it was already demonstrated that so far inaccessible oil deposits can be exploited (BFE, 2003; Jopp, 2003; Randel, 2003; VACI, 2006).

The application of VIPs for insulation of buildings is a matter of scenario analysis. It is expected that there will be further developments. Thereby, transparent VIPs that are covered with a transparent silicon oxide layer and filled with aerogels or nanogels will play an integral role (BINE, 2004, 2011).

4.3.3.5 Polycarbonates for automotive glazing

At present, the use of glass for automotive glazing is established praxis. Disadvantages of glass are its heavy weight and it can be destroyed easily. Following several years of research on innovative glass substitutes, polycarbonate is conducted. But polycarbonate has the attribute that it seasons under ultraviolet irradiation. In order to obviate this process different nanoscale pigments are added. The nanoscale pigments are transparent and do not disturb the view through the window.

Another problem of polycarbonate is scratch sensitivity, but it can be solved with the help of already proven coating processes. Polycarbonate is, therefore, already used for moulded parts such as automotive headlight covers. Glazing is, however, exposed to much higher and dynamic impacts. Through the use of innovative coating technology this level of quality shall be reached (Jopp, 2006). A major advantage of glazing based on polycarbonate in comparison to conventional glass is the weight saving of around 50%. For an average automobile weight savings of 20 kg up to 40 kg can be achieved (Exatec, 2008; McCullough, 2011; Summerer, 2003).

4.3.3.6 Nano-lacquers

The possible application of nano-lacquers in automotives has been analysed since several years. Automobile manufacturers do not put emphasis on potential energy savings, but they perceive nano-lacquers as a competitive edge and as a possibility to increase customer value (NanotechReport, 2003).

Steinfeldt *et al.* (2004, 2007) anaylsed the energy saving potential of nano-lacquers compared to conventional lacquer systems, for example, one-/two-component clear lacquer (1-/2-CCL), water-based clear lacquer (WCL), and powder-based clear lacquer (PCL). The analysis covers the total life cycle of the product lacquer from raw material extraction, base material production, lacquer production, and lacquer application to usage phase.

It was revealed that nano-lacquers can be processed with conventional lacquer technologies and another advantage is that several pre-treatments are no longer necessary when nano-lacquer is applied.

The reason for the energy reduction is down to the considerably lower amount of lacquer required. In addition, the reduced amount of lacquer used reduces the overall weight of an automobile, which results in additional energy savings (dpa, 2009; ntcgmbh, 2008; Steinfeldt *et al.*, 2004, 2007).

4.3.3.7 Nanocatalysts

Nanotechnology used in refineries can affect energy conversion in different fields. The performance of crackers is satisfactory in refining processes since about 20% of crude oil cannot be exploited. The use of nanotechnological catalysts shows a better separating power (selectivity), which makes an increase in production of 2% possible. In addition, nano-based catalysts have the potential to lower the temperature level of refining processes (Boeing, 2003; LANL, 2006). Resulting from the possible widespread use in chemical processes, the use of catalysts based on nanotechnology shows great potential.

According to LANL (2006), the saving potential of lowering process temperatures for USA is about 22 PJ to 90 PJ per year. This means for Germany, based on the German refining capacity in 2010, a possible energy saving potential of 0.9 to 3.6 PJ can be realised.

Nanotechnology catalysts play an important role in the production of chemicals, too. Attributed to the nanostructure of catalysts, the ratio between surface and volume increases, and therewith the amount of so-called reaction centres. LANL (2006) expects that nanotechnology is able to increase activity by 50% of the difference between theoretically possible and actual achievable activity. Similar results were shown in a study (Steinfeldt *et al.*, 2004, 2007) about energetic impact of nano-catalysts for styrene manufacturing. The production of styrene is one of the tenth most important petrochemical processes because it is one of the most important basic chemical product in chemical industry (MPG, 2002). By means of a nanostructured MWCNT, this endothermic process can be transferred into an exothermic process. The production amount can be increased and the operating temperature can be lowered to about 200°C. With the help of a nanoscale catalyst, the overall specific energy consumption for manufacturing 1 kg styrene can be lowered from 6.36 MJ to 3.25 MJ (Steinfeldt *et al.*, 2004, 2007).

4.3.3.8 Nanoparticles in synthetic production

The addition of nanoparticles could change or enhance flowability of materials. For manufacturing processes such as injection moulding this can result in considerable improvements based on the adjustment of new rheological properties. For part-crystalline thermoplastics these possible enhancements are evaluated.

By addition of nanoparticles to polybutylene terephthalate (PBT) it was possible to reduce melt viscosity by around 50% without affecting other properties (mechanical properties, thermal stability, vibration behaviour). The use of nanoparticles results in diverse advantages in processing with injection moulding machines. As a result it is possible to manufacture more thin-walled castings, leading to a reduction of material required.

Moreover, cycle time can be reduced by up to 30%, lower injection pressures are needed, and a reduction of melt temperature is possible. Altogether it is expected to save up to 20% of energy (BASF, 2005; Eipper and Völkel, 2006; Eipper and Stranksy, 2008; Iden, 2010; Weiß and Eipper, 2004).

4.3.3.9 Nanpoarticles in tyre compounds

Generally tyre compounds contain about 30% reinforcing fillers. The properties of rubber compounds such as adhesion, abrasion resistance, or tear resistance are based on these fillers (Degussa, 2006).

The use of soot particles and nanoscale silicic acid is state of the art (Paschen *et al.*, 2004). Through the application of nanotechnological processes it is possible to manufacture new tyre compounds (Velte, 2006). For example, the high surface area of nanostructured soot leads to an increase in interaction between the soot and the rubbers. As a result reinforcement effects in the tyres are intensified (Paschen *et al.*, 2004). With the use of nanoparticles in combination with nanotechnological processes it is supposable to reduce rolling resistance by about 15% to 25% (Velte, 2006). In the case of carbon nanotubes as additives in tyre compounds it seems possible to reduce rolling resistance by about 5% to 25% (Unseld, 2006). In regard to fuel consumption it is assumed that 30% improvement of rolling resistance leads to a reduction in fuel consumption of around 5% (König, 2004).

4.3.3.10 Nano-based coatings to reduce friction

Abrasion and wear-out in engines, bearings, gearings, and other moving parts in mechanical systems is caused by solid body contacts of metallic surfaces. Traditionally it is used in an attempt to generate a separative film with the help of lubricant that should avoid direct contact of the metal surfaces. According to Rewitec (2008) a nano-coating was developed that protects frictional metal surfaces of abrasion and wear-out even if no separative lubricating film can be built up. The coating technology is not based on a modification of lubricant but on the modification of the surface structure of frictional metal parts.

Micro- and nano-silicate particles aggregate under pressure and temperature with metal atoms of the component and build up a wear-resistant metal-silicate coating. Thereby oil and fuel consumption of combustion engines can be reduced (HA, 2007). According to HL (2006) 15% of fuel consumption of modern automobile engines is caused by engine friction. Analyses based on the New European Driving Cycle (NEDC) show that this coating technology can result in a reduction of fuel consumption by around 11% (Thesenvitz *et al.*, 2007). According to Pudenz (2011) an important automotive manufacturer plans to use nano-based coatings for diesel engines in the near future.

4.3.4 *Theoretical Potentials of Nanotechnology*

Based on the detailed analysis of the following nano-based applications, the theoretical potential for applications in the energy sector will be estimated:

- Antireflective coatings
  - Solar thermal
  - CSP
  - Photovoltaics
- New photovoltaic technologies
  - Thin-film solar cells
  - Dye solar cells
  - Organic solar cells
  - Staple solar cells
  - Quantum dot solar cells
- Fuel cell heating units
- Fuel cell vehicles
- Ceroxide
- Nano-based membranes for CCS
- TEGs
  - Automotives
  - Power plants
- Supercapacitors in hybrid buses
- LEDs for multiple-coach lighting
- LEDs in automotive lighting
- OLEDs and LEDs for multiple-coach lighting
- OLED displays
- UHPC
- Insulation with VIPs
- Polycarbonates for automotive glazing
- Nano-lacquers
  - Manufacturing and application
  - Operation
- Styrene manufacturing
- Nanoparticles in synthetic production
- Nanoparticles in tyre compounds
- Nano-based coatings to reduce friction

For estimation of theoretical potential it is assumed that the nano-based innovations have a market penetration of 100%. In regard to efficiencies and technological parameters it is expected that the today on laboratory scale possible values are realised. Table 4.7 summarises the results.

It can be concluded that the 26 nanotechnological applications analysed have theoretically the potential to save 1876 PJ and to generate an additional energy of 404 PJ. The main contributors to generating additional energy are new photovoltaics. They show the potential for generating more than 352 PJ per year, whereas the application of fuel cells in the automotive sector has a major potential for energy savings (630 PJ), and so also do the application of nano-based membranes for CCS (408 PJ), the use of LEDs and OLEDs for multiple-coach lighting (207 PJ) and the use of so-called fuel cell heating units (164 PJ) for energy supply in households. Other major savings are achieved through the use of nano-based coatings to reduce friction in the traffic sector (163 PJ) and the insulation of buildings with nano-based insulation materials (120 PJ).

As shown in Table 4.8 the traffic sector shows the highest theoretical potential for energy savings (916 PJ), followed by the energy generation and conversion sector, with an energy saving potential of 409 PJ and a potential for additional energy generation of around 404 PJ. Households have the potential to save up to 347 PJ. Industry comes last place with 56 PJ of possible theoretical potentials for energy savings.

Table 4.9 shows the comparison between the theoretical potential and reference values (average from 2001 to 2005) according to EU (2006) and BMWi (2011). It becomes obvious that nanotechnology could save around 34% of the energy demand in the traffic sector. In regard to energy generation the theoretical potential could cover almost 19% of electricity generation in Germany.

In order to get a better feeling about the possible impact of nanotechnology on the German energy sector and for possible future savings the following chapter describes the development of scenario and sensitivity analyses based on the estimated theoretical potential.

Table 4.7 Theoretical potential of nano-based application along the energy value chain

		Theoretical potential		
Application		Energy generation	Energy conversion, storage, and distribution	Energy usage
		Additional generation (PJ)	Savings (PJ)	
Antireflective coatings	Solar thermal	2.9		
	CSP	0.8		
	Photovoltaics	0.7		
New photovoltaic technologies	Thin-film solar cells	195.1		
	Dye solar cells	81.1		
	Organic solar cells	29.5		
	Staple solar cells	24.1		
	Quantum dot solar cells	22.3		
Fuel cell heating units			164.5	
Fuel cell vehicles			630.0	
Ceroxide			62.6	
Nano-based membranes for CCS			408.6	
TEGs	Automotives			2.1
	Power plants	47.4		
Supercapacitors in hybrid buses				2.8
LEDs for multiple-coach lighting				134.8
LEDs in automotive lighting				5.0
OLEDs and LEDs for multiple-coach lighting				206.7
OLED displays				33.7
UHPC				17.9
Insulation with VIPs				119.6
Polycarbonates for automotive glazing				1.6
Nano-lacquers	Manufacturing and application			1.0
	Operation			0.2
Styrene manufacturing				7.4

(contd.)

Table 4.7   (*contd.*)

Application	Energy generation	Energy conversion, storage, and distribution	Energy usage
	Additional generation (PJ)	**Savings (PJ)**	
Nanoparticles in synthetic production			0.4
Nanoparticles in tyre compounds			49.2
Nano-based coatings to reduce friction			163.0

(Header spanning: **Theoretical potential**)

Table 4.8   Theoretical potential of nanotechnology for different sectors

	Generation / conversion	Industry	Tertiary sector	Households	Traffic
Energy savings (PJ)	409	56	148	317	916
Additional energy generation (PJ)	404				

(Header spanning: **Theoretical potential**)

Table 4.9   Theoretical potential of nanotechnology for different sectors

Sector	Theoretical potential		Reference value 2001–2005 (according to EU [2006] and BMWi [2011])
	(PJ)	(%)	(PJ)
Additional energy generation	404	18.6	2171
Industry	56	2.2	2485
Tertiary sector	148	9.9	1498
Households	347	12.9	2697
Traffic	916	34.8	2635

Bibliography

Aigle, T. and Jörissen, L. (2003). "Grundlagen der Brennstoffzellentechnik." Weiterbildungszentrum Brennstoffzelle Ulm e.V.

BASF. (2005). "Label Eco-Efficiency Analysis Ultradur® High Speed." BASF Aktiengesellschaft, Ludwigshafen.

BDZ. (2010). "Zahlen und Daten." Bundesverband der Deutschen Zementindustrie e.V., Berlin.

BDZ. (2011). "Zahlen und Daten." Bundesverband der Deutschen Zementindustrie e.V., Berlin.

BFE. (2003). Technologie-Monitoring, Bundesamt für Energie, Bern.

BG-ETEM. (2011). Allgemeinbeleuchtung - in Zukunft LED?, Gesundheit und Sicherheit, Tag für Tag, Berufsgenossenschaft Energie Textil Elektro Medienerzeugnisse.

BINE. (2001). Vakuumdämmung, Projekt-Info 4/01, BINE Informationsdienst des Bundesministerium für Wirtschaft und Technologie, Bonn.

BINE. (2004). "Vakuum-Isolation in Fassadenelementen." BINE Informationsdienst, Projektinfo 08/04.

BINE. (2011). "Dämmen durch Vakuum - Hocheffizienter Wärmeschutz für Gebäudehülle und Fenster." BINE Informationsdienst, Themeninfo I/2011.

Birnbaum, U., Bongartz, R., Linssen, J., Markewitz, P., Vögele, S. (2010). Gemeinschaft STE Research Report - Energietechnologien 2050 - Schwerpunkte für Forschung und Entwicklung, Fossil basierte Kraftwerkstechnologien, Wärmetransport, Brennstoffzellen, Jülich: Forschungszentrum Jülich, Institut für Energieforschung, Systemforschung und Technologische Entwicklung (IEF-STE).

Blum, L. and Peters, R. (2002). "Zukünftiger Einsatz der Brennstoffzelle in der Hausenergieversorgung." Institut für Energieverfahrenstechnik, Forschungszentrum Jülich.

BMBF. (2004). Nanotechnologie: Innovationen für die Welt von morgen, Bundesministerium für Bildung und Forschung, Bonn, Berlin.

BMBF. (2006). Zukunftstechnologien: Von der Idee zur Anwendung, Bundesministerium für Bildung und Forschung, Bonn, Berlin, Jülich, Düsseldorf, Karlsruhe, Köln.

BMBF. (2010). Leuchtdioden (LED) - das Licht der Zukunft, Bundesministerium für Bildung und Forschung, Berlin.

BMWi (2011). Zahlen und Fakten: Energiedaten: Nationale und Internationale Entwicklung, Stand: 07.12.2011, Bundesministerium für Wirtschaft und Technologie, Berlin.

Boeing, N. (2003). "Nanomaterialien zerlegen Toilettendünste - und knacken Rohöl." www.morgenwelt.de.

Brand, L., Eickenbusch, H., Hoffknecht, A., Krauß, O., Zweck, A., and Pohle, D. (2007). Innovations- und Marktpotential neuer Werkstoffe: Monitoringbericht 2007, Zukünftige Technologien Consulting der VDI Technologiezentrum GmbH, Düsseldorf.

BSI. (2007). Nanotechnologie, Bundesamt für Sicherheit in der Informationstechnik, Bonn.

Bullis, K. (2011). Ultracapacitors to Boost the Range of Electric Cars, Technology review, published by MIT.

Bullis, K. and Wirth, R. (2006). "Fantastische Farbenspiele." Technology Review, 2006, 7, S. 84–86.

Degussa. (2006). "Mehr Freude am Sparen: Innovative Füllstoffe für neue Autoreifen." Degussa AG, heute Evonik Industries.

DFG. (2004). Nachhaltiges Bauen mit Ultra-Hochfestem Beton zur Leistungssteigerung, Umweltentlastung und Kostensenkung im Betonbau, DFG-Schwerpunktprogramm (Deutsche Forschungsgemeinschaft).

dpa (2009). Nanolack bietet Hightech - Schutz für das Auto, BerlinOnline Stadtportal GmbH & Co. KG.

Edwards, S. A. (2006). The Nanotech Pioneers: Where are they taking us?, Wiley, Weinheim.

Eipper, A. and Völkel, M. (2006). "Filigrane Bauteile wirtschaftlich herstellen." Kunststoffe - Nanotechnologie (11/2006).

Eipper, A. and Stransky, R. (2008). Kleine Teilchen - großer Effekt, Kunststoffe, Carl Hanser Verlag, München.

ETAP. (2007). "Nanotechnology brings new Solutions for Carbon Capture." Environmental Technologies Action Plan (ETAP).

EU. (2006). "Richtlinie 2006/32/EG des Europäischen Parlaments und des Rates vom 5. April 2006 über Endenergieeffizienz und Energiedienstleistungen und zur Aufhebung der Richtlinie 93/76/ EWG des Rates." Amtsblatt der Europäischen Union.

Exatec. (2008). "Using Exatec Polycarbonate Automotive Glazing to Reshape the Future of Automotive Exteriors." Exatec.

Fairley, P. (2008). "Wenn Diesel auf Nanotechnik trifft." Technology Review (13.09.2008).

FGL. (2006). "Lichtforum 40: LED – die neue Lichtquelle." Fördergemeinschaft Gutes Licht, Frankfurt am Main.

Fischle, H. J. (2005). Superkondensatoren, made by WIMA, WIMA Spezialvertrieb elektronischer Bauelemente GmbH & Co.KG, Berlin.

Fröhlich, K. (2007). Projekt – Anwendungspotential der thermoelektrischen Stromerzeugung im Hochtemperaturbereich, Jahresbericht 2007; ETH Zürich, Institut für elektrische Energieübertragung und Hochspannungstechnologie, Zürich.

Green, M. (2008). "Auf dem Weg zur dritten Solarzellen-Generation." Alfons W. Gentner Verlags und der Solarpraxis AG.

Grüne, M., Kernchen, R., Kohlhoff, J., Kretschmer, T., Luther, W., Neupert, U., Notthoff, C., Reschke, R., Wessel, H., and Zach, H.-G. (2005). Nanotechnologie: Grundlagen und Anwendungen, Fraunhofer Institut für Naturwissenschaftlich-Technische Trendanalysen, Euskirchen, Stuttgart.

Gummert, G. and Suttor, W. (2006). Stationäre Brennstoffzellen: Technik und Markt, Müller (C.F.) Heidelberg.

HA. (2007). "REWITEC: Beschichtungslösungen zur Minderung von Reibung und Verschleiß." Hessen-Nanotech NEWS, 4/2007.

Hagemann, I. (2007). Photovoltaik und Architektur - Positiver Imageträger und innovative Technik für das zukunftorientierte Bauen, altbauplus, Aachen - Fachveranstaltung 16. November 2007, Thema: Außenwand spezial".

Hartmann, U. (2006). Faszination Nanotechnologie, Spektrum Akademischer Verlag, München, Heidelberg.

Helsch, G. (2006). "Haftfeste Antireflexschichten auf Borosilicatglas für Solarkollektoren." TU Clausthal Institut für Nichtmetallische Werkstoffe.

HL. (2006). Dokumentation Nano - Hier ist die Zukunft - Hessen im Dialog, Hessische Landesregierung.

Hofmann, T. (2006). "Nanobeschichtungen für Architektur- und Solargläser." Vortrag vom 24.01.2006, VDI Technologiezentrum, Nanotecture 2006 – Anwendungen der Nanotechnologie in Architektur und Bauwesen, Düsseldorf.

Iden, R. (2010). Der Umwelt zuliebe, Kunststoffe Carl Hanser Verlag, München.

IHK-Braunschweig. (2006). "Haftfeste Antireflexschicht für Borosilikat-Glasrohre in solarthermischen Großkraftwerken." Braunschweig.

ISE. (2006). Jahresbericht 2005: Leistungen und Ergebnisse, Fraunhofer-Gesellschaft Institut für Solare Energiesysteme ISE.

ISE. (2007). "Farbstoff- und Organische Solarzellen." Fraunhofer-Institut für Solare Energiesysteme ISE.

ISE. (2009). Weltrekord: 41,1% Wirkungsgrad für Mehrfachsolarzellen am Fraunhofer-Institut für Solare Energiesysteme ISE.

Jopp, K. (2003). Nanotechnologie – Aufbruch ins Reich der Zwerge, Betriebswirtschaftlicher Verlag Dr. Th. Gabler GWV Fachverlage GmbH Wiesbaden.

Jopp, K. (2006). Nanotechnologie — Aufbruch ins Reich der Zwerge, Betriebswirtschaftlicher Verlag Dr. Th. Gabler GWV Fachverlage GmbH, Wiesbaden.

Jüstel, T. (2006a). Personal message 07.08.2006.

König, U. (2004). "Beiträge der Nanotechnologie für umweltfreundliche Automobile." Symposium Nano meets Umwelttechnik, Fraunhofer IAO Stuttgart, Vortrag vom 02.07.2004.

Kuckelkorn, T. (2006). "Hochtemperaturreceiver für solarthermische Kraftwerke." Symposium Material Innovativ, Universität Bayreuth, Vortrag vom 29.03.2006, Bayreuth.

LANL. (2006). Estimated energy savings and financial impacts of nanomaterials by design on selected applications in the chemical industry, Los Alamos National Laboratory.

LED-Info. (2006). "Grundlagen – LED – Weisslicht LED – weisse LED."

Lewis, N. S. and Argyros, G. L. (2005). "Solar Energy and Nanotechnology;." California Institute of Technology.

Lüdemann, R., Schmidhuber, H., Wirth, H., Brendel, R., Bothe, K., Rech, B., Powalla, M., Oelting, S., and Lang, J. (2005). "Photovoltaik - Innovationen bei Solarzellen und Modulen." FIZ Karslruhe GmbH, Eggenstein-Leopoldshafen.

Luther, W. (2007a). Einsatz von Nanotechnologien in Architektur und Bauwesen, HA Hessen Agentur GmbH, Wiesbaden.

Luther, W. (2008). Einsatz von Nanotechnologien im Energiesektor, HA Hessen Agentur GmbH, Wiesbaden.

McCullough, D. (2011). Innovative und nachhaltige Lösungen für globale Megatrends, kunststoffFORUM, Wolfsburg.

MPG. (2002). "Nanozwiebeln bringen Styrol- Synthese auf Trab." Max-Planck Gesellschaft.

Nanoforum. (2004). Nanotechnology helps solve the world's energy problems, Nanoforum.org, European Nanotechnology Gateway.

NanotechReport. (2003). "Nano gibt Gas." Nanotech Report Ausgabe 09, 2003.

ntcgmbh. (2008). "Nanotechnologie." Nano TechCoatings GmbH.

ObservatoryNano. (2008). "Nanocomposites."

Oey, C. C., Djurisic, A. B., Wang, H., Man, K. K. Y., Chan, W. K., Xie, M. H., Leung, Y. H., Pandey, A., Nunzi, J.-M., and Chui, P. C. (2006). "Polymer-TiO_2 solar cells: TiO_2 interconnected network for improved cell performance." *Nanotechnology*, 17, 2006, 3, S. 7006–7013.

OIDA. (2002). Light Emitting Diodes (LEDs) for General illumination: An OIDA Technology Roadmap Update 2002, Optoelectronics Industry Development Association, Washington DC.

OLED100. (2009). "OLED100.eu - Project Report Three aesthetical perception case studies."

OLED100. (2011). Organic LED lighting in European dimensions, Homepage: http://www.oled100.eu/homepage.asp.

OLLA. (2009). "OLLA Project Report - Final Activity Report."

Oxonica. (2005a). "Envirox: Fuel Economy, Stand: 2005." Oxonica Energy, Kidlington.

Oxonica. (2005b). "Envirox Technical Notes: Catalytic Function of Cerium Oxide, Stand: 2005." Oxonica Energy, Kidlington.

Oxonica. (2007). ENVIROX™ Case Study - Stagecoach 2007, Oxonica.

Oxonica. (2008). Reducing diesel fuel costs with ENVIROX™ fuel additive, Oxonica.

Paddon, P. and Bernhard, M. (2008). "Sonnige Aussichten für Solarzellen – Virtuelles Prototyping von Nanostrukturen." Zeitschrift für Laserphotonik.

Paschen, H., Coenen, C., Fleischer, T., Grünwald, R., Oertel, D., and Revermann, C. (2004). Nanotechnologie - Forschung, Entwicklung, Anwendung, Springer Verlag, Berlin, Heidelberg, New York.

Patel, P. (2011). The First Full-Color Display with Quantum Dots, Technology Review, published by MIT (2011).

Pester, W. and Trechow, P. (2009). "Bordstrom direkt aus Autoabgas zapfen." VDI-Nachrichten, Nr. 44(30, Oktober 2009).

Pudenz, K. (2011). Reibungsverlust minimiert durch Nanoslide-Technologie, ATZ online, Springer Automotive Media, Springer Fachmedien Wiesbaden GmbH, Wiesbaden.

Quaschning, V. (2008). Erneuerbare Energien und Klimaschutz: Hintergründe- Techniken-Anlagenplanung-Wirtschaftlichkeit, Carl Hanser Verlag, München.

Radgen, P., Cremer, C., Warkentin, S., Gerling, P., May, F., and Knopf, S. (2006). Verfahren zur CO_2- Abscheidung und Speicherung Abschlussbericht, Umweltbundesamt.

Randel, P. (2003). "Nanoporöse Dämmstoffe auf Basis Fumed Silica" Vortrag vom 10.07.2003, 1. Fachtagung VIP-Bau, Rostock-Warnemünde.

Reisch, M. (2007). Halbleiter-Bauelemente, Springer Verlag, Berlin.

Rewitec. (2008). "Nanobeschichtung." REWITEC.

RWTH. (2008). "Verbundprojekt Oxycoal- AC - Entwicklung eines CO_2-emissionsfreien Kraftwerksprozesses." RWTH Aachen, Aachener Verfahrenstechnik (AVT).

Steinfeldt, M., Gleich, A. v., Petschow, U., Haum, R., Chudoba, T., and Haubold, S. (2004). Nachhaltigkeitseffekte durch Herstellung und Anwendung nanotechnologischer Produkte, Institut für ökologische Wirtschaftsforschung (IÖW) gGmbH.

Steinfeldt, M., Gleich, A. V., Petschow, U., Haum, R. (2007). Nanotechnologies, Hazards and Ressource Efficiency - A Three-Tiered Approach to Assessing the Implications of Nanotechnology and Influencing its Development, Berlin, Heidelberg: Springer Verlag.

Stobbe, L., Nissen, N. F., Proske, M., Middendorf, A., Schlomann, B., Friedewald, M., Georgieff, P., and Leimbach, T. (2009). "Abschätzung des Energiebedarfs der weiteren Entwicklung der Informationsgesellschaft." Abschlussbericht an das Bundesministerium für Wirtschaft und Technologie, Karlsruhe.

Summerer. (2003). "Exatec - Technologie-Plattform für Kfz-Verscheibungssysteme." Summerer Technologies.

Texocon. (2008). Einführung in die Nanotechnologie, Texocon GmbH, Essen.

Thesenvitz, M., Mohn, B., and Bohl, B. (2007). Prüfbericht: Nanoprotect, Fachhochschule Frankfurt am Main, Labor für Kraftfahrzeugtechnik, Frankfurt am Main.

Tolbert, S. (2008). Improved photovoltaic efficiency in semiconducting polymer/fullerene solar cells through control of fullerene self-assembly and stacking; Research: Project at the University of California in Los Angeles; Reference: UCLA Case No. 2008–662.

UH. (2008). "Umwandlung von Wärme in Elektrische Energie." Pressedienst Universität Hamburg.

Uhlig, R. (2007). Anti-Reflex-Schicht für transparente Hochtemperatur Receiver-Abdeckungen (ARTRANS): Abschlussbericht ARTRANS 2007.

Unseld, K. (2006). Personal message 10.07.2006.

VACI. (2006). "European Research Project VACI on Vacuum Insulation."

va-Q-tec. (2009). "Produktdatenblatt (va-Q-vip B), Stand: 6/2009." va-Q-tec AG, Würzburg.

VDI-TZ. (2006). future technologies update, Verein Deutscher Ingenieure - Technologiezentrum, Düsseldorf.

VDZ. (2007). "Umweltdaten der deutschen Zementindustrie 2007." Verein Deutscher Zementwerke e.V., Forschungsinstitut der Zementindustrie.

Velte, G. (2006). Personal message 03.07.2006.

Viebahn, P., Esken, A., Höller, S., Luhmann, J., Pietzner, K., Vallentin, D., Dietrich, L., Nitsch, J. (2010). RECCS plus Regenerative Energien (RE) im Vergleich mit CO_2-Abtrennung und - Ablagerung (CCS) - Update und Erweiterung der RECCS-Studie, Forschungsvorhaben gefördert vom Bundesministerium für Umwelt, Naturschutz und Reaktorsicherheit, Wuppertal.

vip-bau. (2008). Vip-Bau: Vakuumisolationspaneele, Bayerisches Zentrum für Angewandte Energieforschung e. V. (ZAE Bayern), Würzburg

Weiß, C. and Eipper, A. (2004). "Neues leicht fließendes PBT durch Nanotechnologie." *Plastverarbeiter*, 55. Jahrg. (Nr. 10), S. 178–180.

Wevers, M., Wechsler, D., and Zweck, A. (2002). Nanobiotechnologie I: Grundlagen und technische Anwendungen molekularer, funktionaler Biosysteme, Zukünftige Technologien-Consulting (ZTC) des VDI-Technologiezentrums, Düsseldorf.

WFU. (2007). "Plastic solar cell efficiency breaks record at WFU nanotechnology center." Wake Forest University, News Service (18.04.2007).

Yersin, H. (2010). "Neue OLED-Technologien für Flachbildschirme und Beleuchtung." Universität Regensburg, Institut für Physikalische Chemie, Regensburg.

4.4 Scenario and Sensitivity Analyses for Impacts of Nanotechnological Applications

Jochen Lambauer, Dr. Ulrich Fahl, and
Prof. Dr. Alfred Voß
*Institut für Energiewirtschaft
und Rationelle Energieanwendung (IER),
Universität Stuttgart, Germany*

In the following paragraphs the possible impacts of nanotechnological applications are evaluated by scenario and sensitivity calculations. The description is again based on the energy value chain (energy sources and conversion, energy storage and distribution, as well as energy usage). Due to uncertainties in regard to the future development of nanotechnological applications, three different scenario and sensitivity calculations are developed. The expected development is basis of the scenario calculation (Nano scenario). In order to get a possible corridor of results, a conservative (Nano–) and a progressive (Nano+) development is calculated where parameters such as market penetration and prevalence rate are varied by ±25%, respectively (sensitivity analysis).

In the field of energy generation, possible effects of nanotechnology on the use of solar energy, antireflective coatings for photovoltaics, solar panels, and CSP plants are analysed. In the field of photovoltaics, new technologies, for example, thin-film solar cells, organic polymer solar cells, stack solar cells, and quantum dot solar cells are evaluated. In the field of energy conversion, the application of nanotechnology for fuel cell applications (stationary fuel cells for supply of space heating and generation of electricity for buildings as well as fuel cell vehicles), possible effects of fuel additives (ceroxide) on combustion processes, and the extensive topic of nanomembranes (CCS) are analysed. As there has been substantial progress in the development of TEGs recently, their possible application in power plants and in automotives is evaluated. In the field of energy storage, the application of supercapacitors in hybrid buses is analysed in detail. In the field of energy usage, several nanotechnological applications are evaluated. In regard to illumination, the use of LEDs and OLEDs for multiple-coach lighting as well as the application of LEDs in automotive lighting are a matter of analysis. Furthermore the use of OLED displays is part of the scenario and sensitivity calculations. In the field of construction, the use of UHPC and insulation with VIP is investigated. In addition the following nanotechnological applications and processes are analysed: nano-lacquers (manufacturing, application, and operation), styrene manufacturing, nanoparticles in synthetic production, and tyre compounds, as well as nano-based coatings to reduce friction.

4.4.1 *Energy Sources and Conversion*

In the field of energy sources and conversion, different, current but also new innovations and applications based on nanotechnology are investigated on the basis of scenario and sensitivity analyses. With the help of nanotechnology, current applications can be improved and new innovations become economical and technological ready for entering the market.

4.4.1.1 Solar heat and photovoltaics

By the utilisation of solar energy, different nanotechnological applications are analysed. On the one hand the impact of surface refinement (antireflective coatings) for solar panels, CSP plants and photovoltaic cells is analysed in regard to current technologies. On the other hand, in the field of photovoltaics, new technologies, for example, thin-film solar cells, organic polymer solar cells, stack solar cells, quantum dot solar cells, and dye solar cells, are investigated.

4.4.1.1.1 Antireflective coatings In Section 4.3.1.1.1 the basics about nanotechnological antireflective coatings and their possible application in photovoltaics, solar panels, and CSP plants are described.

By the use of antireflective coatings, the losses (around 10%) due to reflections can be reduced and therefore the efficiency of solar energy can be enhanced. It is assumed that for photovoltaics, the power output can be increased by up to 4 percentage points. For solar panels (heating and tap water), an increase of up to 6 percentage points is predicted (Hofmann, 2006). For the Nano scenario it is therefore assumed that the efficiency of solar panels increases by 5.5 percentage points and the efficiency of photovoltaics increases by 3.5 percentage points. For CSP plants it is assumed that the energy output can be enhanced by 4 percentage points. The basis for the calculation is the projection of the future development of solar energy use (solar thermal energy, photovoltaics, CSP plants [based on the idea of a European electricity network for renewable energies]) according to (BMWi, 2011) and Nitsch *et al.* (2010). Furthermore a lifetime of 20 years is postulated for solar panels and photovoltaics. For market penetration it is

Table 4.10   Additional energy generation due to antireflective coatings

Application	Unit	2010	2015	2020	2025	2030
Nano −						
Solar thermal	Additional heat generation (PJ)	0.0	0.1	0.8	1.5	2.1
Photovoltaics	Additional electricity generation (PJ)	0.0	0.1	0.3	0.2	0.5
CSP	Additional electricity generation (PJ)	0.0	0.0	0.0	0.2	0.5
Nano						
Solar thermal	Additional heat generation (PJ)	0.0	0.1	1.0	2.0	2.8
Photovoltaics	Additional electricity generation (PJ)	0.0	0.1	0.4	0.3	0.7
CSP	Additional electricity generation (PJ)	0.0	0.0	0.0	0.4	0.8
Nano +						
Solar thermal	Additional heat generation (PJ)	0.0	0.2	1.3	2.5	2.9
Photovoltaics	Additional electricity generation (PJ)	0.0	0.1	0.5	0.4	0.7
CSP	Additional electricity generation (PJ)	0.0	0.0	0.0	0.6	1.3

assumed that the market share of antireflective coatings starts from 10% in 2015 and will increase up to 95% in 2030. In the case of CSP plants, the calculation expects that the enhancement in efficiency of 4 percentage points starts in 2020. Table 4.10 shows the possible additional energy generation due to the use of antireflective coatings for the different scenario and sensitivity calculations. It becomes obvious that till 2030, there is insignificant impact of antireflective coatings on energy generation.

4.4.1.1.2 New photovoltaic technologies   Next to the improvement of existing photovoltaic cells made of mainly silicon, emphasis is put on developing new and innovative photovoltaic cells (Jopp, 2006). Examples for new cell types are thin-film solar cells or dye solar cells. For the development of these new technologies, nanotechnology plays an important role. Therefore, the following new technologies are analysed in detail in the scenario calculations:

- Thin-film solar cells (a-Si, CdTe, Cis/CIGS)
- Dye solar cells
- Organic polymer cells
- Stapel solar cells
- Quantum dot solar cells

In regard to the diffusion rate and the possible fields of applications of these new cells, building roofs and facades of buildings are regarded as possible application areas. Some more estimations for possible areas can be found in literature. Bernreuter (2002) summarises different estimations. In the scope of this book estimations based on (Kaltschmitt *et al.*, 2006) are used as basis for the scenario calculations. The potential areas are determined, according to (Kaltschmitt *et al.*, 2006), as from existing roofs, partly from building facades and partly from agricultural areas that are not used or needed for food production. Potential usable areas such as railway roofing, parking area roofing, arcades, or roofed entrances are assumed to be unsuitable for energy generation.

The usable areas on roofs for photovoltaic modules is determined from building inventory, average building roof area, style of roof, and slope of roof, as well as structural restrictions in regard to roof lights or funnels and solar technological restrictions such as shadowing effects and safety clearances. In addition the possible roof area for photovoltaics is in competition with the use by solar panels. As a result the usable roof area is determined to be 838 million m^2. Besides roof areas, about 200 million m^2 of facades are possible areas for using photovoltaics. Other possible areas such as agricultural areas that are no longer used for food production are not included in the calculation. In conclusion, for the scenario calculations, according to Kaltschmitt *et al.* (2006), the following possible areas for photovoltaics are used:

- Roof areas: 419 million m^2 (in regard to the competition with solar thermal collectors)
- Facade areas: 200 million m^2

Table 4.11   Efficiencies and market penetration for thin-film solar cells

| | Market penetration and efficiencies η (%) | | | | | | | | |
| | a-Si thin-film solar cells | | | CdTe thin-film solar cells | | | CIGS thin-film solar cells | | |
Year	Roofs	Facades	η	Roofs	Facades	η	Roofs	Facades	η
2010	0.0	0.0	6.0	0.0	0.0	8.0	0.0	0.0	11.8
2015	1.3	1.3	8.3	0.6	0.8	10.1	1.3	1.3	13.9
2020	2.5	2.5	10.5	1.3	1.5	12.3	2.5	2.5	15.9
2025	3.8	3.8	12.8	1.9	2.3	14.4	3.8	3.8	18.0
2030	5.0	5.0	15.0	2.5	3.0	16.5	5.0	5.0	20.0

In regard to efficiencies of photovoltaics, it is postulated that they will further develop and will reach the values that are currently achieved on a laboratory scale. The calculation of the yearly electricity output (see Eqs. 4.1 and 4.2) is based on Quaschning (2008). Losses due to inclination and the performance ratio are neglected. The average global radiation in Germany is assumed to be 1,000 kWh/(m²*a). Efficiencies and market penetration for the Nano scenario are stated in Tables 4.11 and 4.12.

$$E_{elect.} = \frac{H_{solar} * f_{incl.} * P_{MPP} * PR}{1\frac{kW}{m^2}} \quad (4.1)$$

with

$$P_{MPP} = A * \eta * 1\frac{kW}{m^2} \quad (4.2)$$

$E_{elect.}$	yearly electricity output	– kWh/a
H_{solar}	average global radiation	– kWh/(m²*a)
F_{incl}	gains and losses due to inclination	
P_{MPP}	nominal power	– kW$_p$
PR	Performance ratio	
A	usable area	– m²
η	efficiency	

On the basis of the development of market penetration and future efficiencies of the different photovoltaic technologies, Table 4.13 summarises the possible amount of electricity generated by

Table 4.12   Efficiencies and market penetration for new photovoltaic technologies

		Market penetration and efficiencies η (%)				
		2010	2015	2020	2025	2030
Dye solar cells	Roofs	0.0	1.8	3.5	5.3	7.0
	Facades	0.0	1.8	3.5	5.3	7.0
	η	5.0	7.0	9.0	11.0	13.0
Organic polymer cells	Roofs	0.0	0.6	1.3	1.9	2.5
	Facades	0.0	1.0	2.0	3.0	4.0
	η	5.0	6.3	7.5	8.8	10.0
Staple solar cells	Roofs	0.0	0.5	1.0	1.5	2.0
	Facades	0.0	0.0	0.0	0.0	0.0
	η	10.0	12.5	15.0	17.5	20.0
Quantum dot solar cells	Roofs	0.0	0.3	0.5	0.8	1.0
	Facades	0.0	0.0	0.0	0.0	0.0
	η	10.0	13.8	17.5	21.3	25.0

new photovoltaic technologies till 2030 for the different scenarios. The results show that nanotechnological-based improvements of thin-film solar cells have the highest potential for additional electricity generation in future. Most other technologies are still in development.

Therefore it is possible that due to further enhancements in efficiencies, in future, they could play an even more important part in comparison with the actual projection.

4.4.1.2 Fuel cells

Fuel cells can be used to produce both heat and electricity at the same time. This is called combined heat and power (CHP). For a stationary use of fuel cells, the so-called fuel cell heating units can be used. They are made of a fuel cell, an additional gas-powered heater, and a heat storage tank. When using a fuel cell heating unit next to the production of the needed heat, electricity is produced and can be eventually used for one's own consumption or fed into the electricity grid. Therefore each fuel cell heating unit is linked to the public electricity grid. Depending on the operating mode and the actual electricity demand, electricity is fed into the grid or current is purchased. For heat and electricity production in stationary fuel

Table 4.13   Additional electricity generation due to new photo-voltaic technologies

Additional electricity generation (PJ)	2010	2015	2020	2025	2030
Nano −					
a-Si thin-film solar cells	0.0	1.3	3.3	6.0	9.4
CdTe thin-film solar cells	0.0	0.8	2.0	3.6	5.5
CIGS thin-film solar cells	0.0	2.2	5.0	8.4	12.5
Dye solar cells	0.0	1.5	3.9	7.2	11.4
Organic solar cells	0.0	0.6	1.4	2.5	3.7
Staple solar cells	0.0	0.5	1.3	2.2	3.4
Quantum dot solar cells	0.0	0.6	1.5	2.7	4.2
Nano					
a-Si thin-film solar cells	0.0	2.3	5.8	10.7	16.7
CdTe thin-film solar cells	0.0	1.5	3.6	6.4	9.8
CIGS thin-film solar cells	0.0	3.9	8.9	15.0	22.3
Dye solar cells	0.0	2.7	7.0	12.9	20.3
Organic solar cells	0.0	1.0	2.5	4.4	6.7
Staple solar cells	0.0	0.9	2.3	4.0	6.0
Quantum dot solar cells	0.0	0.8	1.9	3.6	5.6
Nano +					
a-Si thin-film solar cells	0.0	3.6	9.1	16.6	26.1
CdTe thin-film solar cells	0.0	2.3	5.7	10.0	15.3
CIGS thin-film solar cells	0.0	6.0	13.8	23.4	34.8
Dye solar cells	0.0	4.3	11.0	20.1	31.7
Organic solar cells	0.0	1.6	3.9	6.8	10.4
Staple solar cells	0.0	1.5	3.5	6.2	9.4
Quantum dot solar cells	0.0	1.2	3.0	5.5	8.7

cells, it is technically possible to use, next to hydrogen, fuel gas in the short and the midterm – mainly natural gas (Edel, 2003; Edel, 2006). In CHP mode, efficiencies of more than 90% are possible for stationary fuel cells (IEA, 2005). In order to analyse possible impacts of fuel cell heating units in Germany, the technical potential for fuel cell heating units is first determined. These units can be used in natural gas–supplied areas as a substitute for a conventional gas boiler (reinvestment), or they can replace heating units that are powered with other energy sources (change in energy carriers). Another potential for the use of fuel cell heating units is gas supplied new construction.

In regard to the building structure, it is distinguished between single-family houses, semidetached houses, and multi-family houses. On the basis of different possible capacities of fuel cell heating units, there are almost no properties that are not suitable for the use of fuel cell heating units (Gummert and Suttor, 2006). Therefore the total natural gas–supplied area is used as a potential for these heating units. Of relevance for the calculation of the possible impacts on the energy demand are heating demand, hot water demand and electricity requirement.

Therefore characteristic demand values of reference buildings are used. For market penetration it is divided between single-family houses, semidetached houses, and multi-family houses. According to Edel (2006), for fuel cell heating units, a faster diffusion in single-family houses and semidetached houses can be expected. This is due to problems concerning the split-up of the produced energy to the different housing units in multi-family houses. Further it is divided between new construction and existing buildings (see Tables 4.14 and 4.15). In regard to the lifetime of heating units (20 years for boilers and peripheral components) and the prevalence rate for fuel cell units, the number of fuel cell units that will be used can be calculated on a yearly basis. With all these parameters the impact of fuel cell heating units on the energy demand in Germany can be calculated.

Table 4.14   Market penetration for fuel cell heating units in new constructions

Market penetration (%)	2010	2015	2020	2025	2030
Nano −					
Single-family houses	0.0	6.8	20.3	22.5	24.0
Semidetached housed	0.0	6.8	20.3	22.5	22.5
Multi-family houses	0.0	1.1	3.4	10.1	11.3
Nano					
Single-family houses	0.0	9.0	27.0	30.0	32.0
Semidetached housed	0.0	9.0	27.0	30.0	30.0
Multi-family houses	0.0	1.5	4.5	13.5	15.0
Nano +					
Single-family houses	0.0	11.3	33.8	37.5	40.0
Semidetached housed	0.0	11.3	33.8	37.5	37.5
Multi-family houses	0.0	1.9	5.6	16.9	18.8

Table 4.15   Market penetration for fuel cell heating units in existing constructions

Market penetration (%)	2010	2015	2020	2025	2030
Nano −					
Single-family houses	0.0	2.3	6.8	20.3	22.5
Semidetached housed	0.0	2.3	6.8	20.3	22.5
Multi-family houses	0.0	1.1	3.4	10.1	11.3
Nano					
Single-family houses	0.0	3.0	9.0	27.0	30.0
Semidetached housed	0.0	3.0	9.0	27.0	30.0
Multi-family houses	0.0	1.5	4.5	13.5	15.0
Nano +					
Single-family houses	0.0	3.8	11.3	33.8	37.5
Semidetached housed	0.0	3.8	11.3	33.8	37.5
Multi-family houses	0.0	1.9	5.6	16.9	18.8

Table 4.16   Energy savings due to the use of fuel cell heating units

Energy savings (PJ)	2010	2015	2020	2025	2030
Nano −	0.0	2.1	9.7	30.8	61.0
Nano	0.0	2.8	12.9	41.1	81.4
Nano +	0.0	3.5	16.1	51.4	101.7

Table 4.16 shows the projection of the possible energy savings on the basis of the use of the fuel cell heating units in German households. It is expected that till 2030, more than 81 PJ could be saved.

The use of fuel cells has been discussed and tested for a long time as a replacement for conventional engines in automotives. According to Strategierat (2006), fuel cell vehicles could lead to an increase in efficiency, ensure reliable energy supplies, and help to reduce CO_2 emissions. Already today nanotechnology is used to improve fuel cells (e.g., nanostructured platinum catalysts), and in future it is expected that it will contribute further to performance enhancement and to cost reduction. For the scenario and sensitivity analyses the technical potential for fuel cells is assumed to be 100% based on all vehicles in Germany. Fuel cell vehicles substitute petrol and diesel engines in equal measure. For reference a fuel cell research vehicle is used. In the reference vehicle a fuel cell (power output

Table 4.17   Energy savings due to the use of fuel cell vehicles

Energy savings (PJ)	2010	2015	2020	2025	2030
Nano −	0.0	0.1	0.7	3.0	7.4
Nano	0.0	0.1	1.0	4.1	10.2
Nano +	0.0	0.2	1.4	5.4	13.3

80 kW) is combined with a Li-ion accumulator (300 V). Hydrogen consumption of the fuel cell vehicle is equal to a fuel consumption of 2.9 l diesel for 100 km (Daimler, 2010a; Geitmann, 2005). Entry of fuel cell vehicles into the mass market is expected to start in 2020 with 0.1% of the total vehicle fleet and will go up to 1% in 2030. Table 4.17 shows the possible impacts of fuel cell vehicles for the different scenario and sensitivity calculations till 2030. Due to the sensitivity calculations the possible energy savings range from 7.4 to 13.3 PJ in 2030.

4.4.1.3 Fuel additives

Basics about possible reduction in fuel consumption by the use of a fuel additive ceroxide are shown in Section 4.3.1.3. On the basis of that, the potential for energy savings is estimated. According to Wilk (2006) ceroxide can be burned technically in every diesel engine. Therefore the technical potential for diesel-driven vehicles is assumed to be 100%. In regard to insecurities concerning the possible use of ceroxide in petrol engines, the technical potential is reduced to 50% and the market entry is delayed for five years. Possible savings in fuel consumption vary between 2% and 9%.

For diesel engines it is assumed that by 2015, already around 10% of all diesel-driven vehicles will use ceroxide as a fuel additive. In 2030 market penetration of the fuel additive ceroxide is determined to be 50% (see Table 4.18).

Table 4.19 shows the results of the scenario and sensitivity calculations. It is expected that ceroxide will reduce fuel consumption till 2030 of around 35 PJ.

Table 4.18   Market penetration for the use of ceroxide as fuel additive

Market penetration (%)	2010	2015	2020	2025	2030
Nano −					
Petrol	0.0	0.0	7.5	15.0	26.3
Diesel	0.0	7.5	15.0	26.3	37.5
Nano					
Petrol	0.0	0.0	10.0	20.0	35.0
Diesel	0.0	10.0	20.0	35.0	50.0
Nano +					
Petrol	0.0	0.0	12.5	25.0	43.8
Diesel	0.0	12.5	25.0	43.8	62.5

Table 4.19   Energy savings due to the use of fuel additives

Energy savings (PJ)	2010	2015	2020	2025	2030
Nano −	0.0	1.0	3.2	6.1	9.5
Nano	0.0	3.9	11.7	22.4	34.8
Nano +	0.0	8.0	24.1	45.7	70.9

4.4.1.4 Nano-based membranes for carbon capture and storage

In Section 4.3.1.4 different possible applications of nano-based membranes are described. In the scope of the scenario and sensitivity analyses, the possible impacts of the use of nanomembranes for carbon capture and storage (CCS) are evaluated. CCS technologies show, owing to an additional energy demand for capturing CO_2, lower efficiencies and thus a higher primary energy demand. But in comparison to chemical proceedings for CCS, membrane technologies show lower losses in efficiency. According to Birnbaum *et al.* (2010) it is assumed that net efficiency of coal-fired power plants with the current state of technology decreases by around 10 to 14 percentage points. This leads to an increase in the own energy demand of the power plant by around 15% to 30%. By further technological development, especially with improvements in gas separation membranes, it is expected that the use of nanomembranes reduces occurring losses in efficiency. It should be possible according to ETAP (2007) to lower the increase of the own energy demand to around 8%. This would result in a decrease of efficiency by around 3 percentage points. According to Birnbaum *et al.* (2010) and Viehbahn *et al.* (2010) losses in efficiency of around 4 to 8 percentage points seem achievable in future.

The calculation is done separately for hard coal, lignite, and gas power plants in regard to their different efficiencies. Furthermore it is therewith possible to integrate a different market penetration for CCS technologies for the different conventional power plant technologies.

It is assumed that CCS technologies are applied to coal power plants earlier than in gas power plants. For hard coal and lignite power plants it is assumed that by 2015, already 2% of the power plants will be equipped with CCS technologies. This value will increase to 70% until 2030. One reason why gas power plants are equipped with CCS after coal and lignite powered plants is, for example, that fuel gas generates less CO_2 emissions and therefore the specific costs for CO_2 separation are higher in comparison to coal-powered power plants. Therefore market penetration for gas power plants start with 5% in 2020 and goes up to 60% in 2030. On the basis of future electricity production, further developments in efficiencies and market penetration of CCS technologies can result in a possible increase in primary energy demand by the use of these technologies. With the help of the calculation, it can be appraised which impacts and possibilities could result from nanotechnology in regard to CCS. Table 4.20 shows the expected savings in primary energy consumption (PEC) due to the use of nano-based membranes for CCS in comparison to current CCS technology with an increase in net efficiency by around 8 percentage points. With the help of nanotechnology, the PEC could be reduced by around 260 PJ in 2030.

Table 4.20   Reduction in PEC due to the application of nano-based membranes for CCS

Reduction in PEC (PJ)	2010	2015	2020	2025	2030
Nano −	0.0	6.4	57.0	109.1	195.2
Nano	0.0	8.5	76.0	145.5	260.2
Nano +	0.0	10.7	95.0	181.9	325.3

4.4.1.5 Thermoelectric generators

According to Fröhlich (2007) efficiencies of TEGs range from around 5% to 10%. By the use of nanostructured materials, a TEG is expected to have efficiency improvements of 50% or even up to 200% (UH, 2008). With an alloy consisting of lead, tellurium, and nanoparticles made of silver and antimony, conversion from heat energy into electricity at a temperature difference of 600°C was possible with an efficiency of 18% (Bachmann, 2007). Therefore application of TEGs in power plants as well as automobiles is conceivable. In the scope of this calculation all conventional power plants are matter of the analysis. The usable waste heat results from the possible efficiencies of the different power plants. It is assumed that only 70% of the occurring waste heat is available for electricity generation by TEGs. As up to now TEGs are not yet available on a large scale the efficiency of the generators is assumed to be 5%, which is far below the possible realised efficiencies according to (Bachmann, 2007).

For market penetration it is optimistically assumed that by 2020, already around 5% of the power plants will be equipped with TEGs. In 2030 the percentage will go further up to 30%. The application of TEGs would increase electricity production and subsequently decrease specific primary energy demand as well as CO_2 emissions. Table 4.21 shows the development of the possible additional energy generation in conventional power plants with the use of TEGs till 2030. According to the results of the scenario and sensitivity calculations, TEGs could produce between 11 PJ and 18 PJ of additional electricity using waste heat of power plants.

TEGs could be applied not only in power plants but also, possibly, in automotives. According to Paschen *et al.* (2004) the application of

Table 4.21   Electricity generation by the use of TEGs in power plants

Electricity generated with TEG (PJ)	2010	2015	2020	2025	2030
Nano −	0.0	0.0	3.1	6.9	10.7
Nano	0.0	0.0	4.1	9.2	14.2
Nano +	0.0	0.0	5.2	11.5	17.8

TEGs on the engine, in the exhaust gas system, or on the catalyst of a vehicle is conceivable. By 2012 TEGs will be used in automotives, according to Pester and Trechow (2009), in order to gain electrical energy to feed into the on-board power supply based on the heat of exhaust gases. It is assumed that TEGs can generate up to 250 W out of the hot exhaust gases of the exhaust gas recirculation. This should result in a decrease of fuel consumption by around 2%.

For the scenario and sensitivity calculations the possible reduction in fuel consumption is varied from 1.1% up to 2%. For petrol- and diesel-driven vehicles, it is assumed that 100% of all new vehicles are a potential for the use of TEGs. Because it can be reckoned with vehicles that are equipped with these generators only after 2012 for the scenario calculations, it is assumed that starting from 2015, TEGs will be an option to reduce fuel consumption. For market penetration it is assumed that in 2020, around 2% of all new vehicles and in 2030 40% of all new vehicles will use TEGs. Table 4.22 shows the possible fuel savings through the use of TEGs in automotives. Till 2030 fuel consumption could be reduced by around 1.1 PJ.

Table 4.22   Fuel savings due to the use TEGs in automotives

Energy savings (PJ)	2010	2015	2020	2025	2030
Nano −	0.0	0.0	0.1	0.5	0.8
Nano	0.0	0.0	0.1	0.7	1.1
Nano +	0.0	0.0	0.2	0.9	1.4

4.4.2 *Energy Storage and Distribution*

Kinetic energy of braking cannot be used in conventional urban buses equipped with mechanical braking systems. When a bus is electrically decelerated, the recaptured braking energy can be stored, for example, in so-called supercapacitors and used again for starting. Thereby fuel consumption, vehicle noise pollution during starting, and emissions can be reduced. On the basis of their technical properties, for this kind of use supercapacitors are superior in comparison to other energy storage devices (MAN, 2005). Therefore the possible impacts of hybrid urban buses equipped with supercapacitors as storage devices are analysed. For the scenario and sensitivity calculations, the technical potential of urban buses in Germany is used and an average lifetime of 12 years is assumed. Average fuel consumption of urban buses is around 1,346 MJ/100 km (Alber, 2006; UBA-CH, 2004). Depending on operation (layout of roads, stops, etc.), savings in fuel consumption of around 20% to 25% are possible. This huge saving potential is dominated by the effect of brake energy recuperation of up to 80% (Fischle, 2005; Kreschl *et al.*, 2005). Similar savings in fuel consumption are possible (according to Daimler, 2009; Daimler, 2010b and Daimler, 2011) with a hybrid urban bus based on the use of a Li-ion accumulator as energy storage. For the scenario and sensitivity calculations the possible savings vary between 20% and 25%. For market penetration it is assumed that in 2015, about 3.5% and in 2030 about 9% of urban buses in Germany will be equipped with hybrid technology. Table 4.23 shows the results, and it can be seen that till 2030, supercapacitors in urban buses could result in fuel savings of around 0.5 PJ till 1.1 PJ, depending on market penetration.

Table 4.23   Fuel savings due to the use of supercapacitors in hybrid buses

Energy savings (PJ)	2010	2015	2020	2025	2030
Nano −	0.0	0.1	0.2	0.3	0.5
Nano	0.0	0.1	0.2	0.5	0.8
Nano +	0.0	0.1	0.3	0.7	1.1

4.4.3 *Energy Use*

4.4.3.1 LED and OLED in illumination

In the field of new and innovative illumination technologies based on nanotechnology, the use of LEDs and OLEDs is analysed in the scenario and sensitivity calculations. Light sources for multiple-coach lighting can be divided into three lamp types: light bulbs, discharge lamps (low- and high-pressure discharge lamps), and LEDs (in future, also OLEDs). Fluorescent lamps as well as compact fluorescent lamps (CFLs) are counted among the group of low-discharge lamps. Generally there are no technical restrictions for the prevalence of new illumination technologies. Given that high-pressure discharge lamps already show high efficiencies (Jüstel, 2006), they are excluded from the substitutable potential of light sources. First the potential for the substitution of conventional lamps by LEDs and OLEDs is determined. In order to get a general comparability of the different single technologies based on the potential for substitution, the energy demand of each illumination technology is specified. In doing so the light amount per life cycle is calculated with the help of the specific lifetime and an averaged luminous flux. In combination with a defined reference light quantity, the number of lamps for each illumination technology is calculated. Consequently, the energy demand for each lamp type can be calculated. The change in energy demand is a result of the energy demand per lamp type multiplied by the possible energy saving per illumination technology. The consideration is done separately for different sectors, industry, the commercial sector, as well as households. Figure 4.5 shows the approach for the analysis of new illumination technologies.

For technical parameters and the future development of these parameters for LEDs and OLEDs, see Tables 4.24 and 4.25 (Jüstel, 2006; OLED100, 2009; OLLA, 2009; Steinfeldt *et al.*, 2004; Steinfeldt *et al.*, 2007). According to OLLA (2009) OLEDs with a lifetime of more than 100,000 hours and with an efficiency or light yield of 50 lm/W are already realised. In order to go into market, light yields of about 100 lm/W are needed.

These technical parameters are the aim of an ongoing research project (see OLED100, 2009, 2011). In the scenario calculations the ban of traditional light bulbs comes into effect in 2012, and until 2015 all remaining light bulbs will gradually be replaced by energy-saving lamps.

Table 4.26 summarises the possible energy savings by the use of LEDs and OLEDs till 2030. According to the scenario and sensitivity calculations, the use of LEDs could save around 115 PJ till 2030. Additional savings of 38 PJ could be achieved through the replacement of conventional illumination by the additional use of

Table 4.24   Technical parameters of conventional illumi-nation technologies

	Unit	Light bulb	CFL
Power input	W	75	15
Lifetime	h	1000	13000
Light yield	lm/W	12	60

Table 4.25   Technical parameters of LEDs and OLEDs

	Unit	2010	2015	2020	2025	2030
			LED			
Power input	W	1	1	1	1	1
Lifetime	h	15000	15000	15000	15000	15000
Light yield	lm/W	30	100	150	200	220
			OLED			
Power input	W	1	1	1	1	1
Lifetime	h	10000	15000	100000	100000	100000
Light yield	lm/W	50	80	100	150	200

Table 4.26   Energy savings by the use of LEDs and OLEDs for illumination

Energy savings (PJ)	2010	2015	2020	2025	2030
		Nano −			
LED	0.0	21.8	32.3	57.4	90.9
LED and OLED	0.0	21.8	36.2	70.8	124.8
		Nano			
LED	0.0	23.0	37.0	70.5	115.2
LED and OLED	0.0	23.0	42.0	86.4	153.1
		Nano +			
LED	0.0	24.2	41.8	83.6	134.3
LED and OLED	0.0	24.2	47.6	101.1	174.5

OLEDs. Altogether between 91 PJ and 175 PJ could be saved by new illumination technologies, such as LEDs and OLEDs, till 2030.

The use of LED is not limited to multiple-coach lighting; it has in fact been utilised as light sources in automotives. Illumination has a share of about 15.7 PJ in the total energy demand of the traffic sector in Germany in 2008. Considering a long lifetime, design flexibility, as well as the possibility to save energy, the use of LEDs in automotive lighting is of great interest (Schiermeister *et al.*, 2003). However, extensive use in headlights is not expected in the short run. For dimmed headlights of a passenger car, which are parking lights, license plate illumination, and rear lights, about 110 W are needed. The fuel consumption by using conventional light bulbs in case of petrol engines is around 0.207 l/100 km and for diesel engines it is about 0.142 l/100 km (BASt, 2006).

The use of LED headlights can reduce energy demand of car illumination by about 39% (see Hamm, 2006). This reduction results in savings in fuel consumption. For petrol engines the fuel consumption for illumination can be reduced to 0.13 l/100 km and for diesel engines to 0.09 l/100 km. The technical potential is assumed to be 100% of all new vehicles. For market penetration it is assumed that by 2015, already 10% of new vehicles will be equipped with LED lighting technology and will increase to about 90% in 2030. Table 4.27 shows the results of the calculations. In 2030 between 3.4 PJ and almost 5.3 PJ could be saved by the use of LEDs in automotive headlights.

Table 4.27   Fuel savings due to the use of LEDs in automotives

Energy savings (PJ)	2010	2015	2020	2025	2030
Nano −	0.0	0.3	1.0	1.9	3.4
Nano	0.0	0.4	1.4	2.6	4.6
Nano +	0.0	0.5	1.7	3.2	5.3

4.4.3.2 New display technologies

The application of OLED displays to substitute conventional display technologies has not yet taken place. Different manufacturers have already presented the first models, but they are still far away from a widespread market entry. OLED displays stand out in regard to their considerably lower energy consumption during operation. Additionally in regard to the whole life cycle (preproduction, production, and operation), OLED displays show energetic advantages in comparison to conventional display technologies, such as CRT or LCD. For the scenario and sensitivity calculations

Table 4.28 Energy demand and power input for different display technologies

Technology	Power input in 2010 (W)	Energy demand (operation) (MJ)	Energy demand in comparison to OLED display (%)
CRT	200	1,990	930
LCD	100	710	332
PDP	100	710	332
OLED	30	214	100

for the substitution of conventional display technologies by OLED displays, at first the potential for substitution is determined. It is assumed that OLED displays can substitute all three evaluated conventional display technologies (CRT, LCD, and PDP). On the basis of assumptions concerning the prevalence rate for OLED displays and the share of displays in regard to the energy demand of the different sectors, the energy demand for each display technology can be calculated. In order to compare the power inputs of the different display technologies, the input is adjusted to the size of the display, according to Steinfeldt *et al.* (2004, 2007). This is based on an average power input of a 50-inch display in 2010 (see Table 4.28). The calculation of the future energy demand of OLED displays takes into account that the efficiency of OLED displays will increase corresponding to further technological development. Energy demands of different display technologies are calculated according to Stobbe *et al.* (2009) for the reference year 2010.

The results of the scenario and sensitivity calculations based on the just-mentioned assumptions are shown in Table 4.29. From around 0.3 PJ in 2015 up to 10 PJ in 2030 could be saved by the use of innovative display technologies in industry, the tertiary sector, and households.

Table 4.29 Energy savings due to the use of OLED displays

Energy savings (PJ)	2010	2015	2020	2025	2030
Nano −	0.0	0.2	2.3	4.7	8.5
Nano	0.0	0.3	2.8	5.7	10.6
Nano +	0.0	0.4	3.6	7.4	13.6

4.4.3.3 Ultra-high-performance concrete

In 2009, German cement manufacturing alone had an energy demand of around 100 PJ (BDZ, 2011). Basic information about UHPC is described in Section 4.3.3.3. As a basis for the calculation, instead of the obstructed amount of cement, the total amount of produced cement is used.

The application of high-performance concrete allows a considerable reduction in the amount of concrete needed and, subsequently, the raw materials and energy required. The possible reduction in energy demand by the application of high-performance concrete in comparison to conventional concrete varies in the scenarios in a range of 10% to 40%. Further input parameters are the amount of cement produced in Germany, the projections of the specific energy demand (heat and electricity) for the manufacturing of cement until 2030, and possible market penetration. For the calculation of the energy demand of the conventional manufacturing process, it is assumed that the efficiency increase of the past can be realised as well in future.

For electrical energy demand, a decrease in specific energy consumption of 0.8% per year and for thermal heat demand a decrease of 1.1% per year is used for the Nano scenario. In order to cover possible uncertainties of future development, the specific energy demand for conventional production varies in a range from 0.4% to 1.2% (electrical energy demand) and 0.9% to 1.6% (thermal energy demand). With the theoretical energy savings and the market penetration (2015: 0.5%; 2030: 2.5%) of UHPC, the possible energy reduction in the construction sector is calculated. Table 4.30 summarises the possible effects through the use of UHPC. It is obvious that till 2030, there is still little impact by the use of UHPC on energy consumption. According to the calculations, between 0.2 and 1.1 PJ could be saved in 2030.

Table 4.30   Energy savings due to the use of UHPC

Energy savings (PJ)	2010	2015	2020	2025	2030
Nano −	0.0	0.0	0.1	0.1	0.2
Nano	0.0	0.1	0.3	0.4	0.4
Nano +	0.0	0.2	0.7	0.9	1.1

4.4.3.4 Insulation with vacuum-insulation panels

In the household sector, energy demand for heating has a share of about 72% (1,833 PJ in 2008) on the total energy demand (BMWi, 2011). Next to modernisation of heating technologies, insulation of buildings can reduce heat demand, and therefore insulation is another great potential for energy savings. Furthermore efficient insulation reduces warming of buildings in summertime, and in this way energy demand for the cooling system can be avoided. According to Jopp (2006) VIPs have brilliant perspectives in the whole construction sector: as insulation material in the building inventory (energetic renovation) as well as in the field of new constructions. For that reason, new construction is also a potential area for the use of VIPs in the scope of the scenario and sensitivity calculations.

The approach for the calculation is shown in Figure 4.6 in chapter 4.1. Having identified the number of housing units that could be equipped with VIP insulation (potential for substitution), as well as the prevalence rate to use VIPs when an insulation is carried out, the number of housing units with VIP insulation is defined on a yearly basis.

In order to compare the impact of VIPs, PS hard foam is used as reference insulation material. On the basis of technical parameters of the different insulation materials, the k-value[6] of a 1 m^2 outer wall surface is calculated.

For the calculation of the transmission heat demand of the outer wall, the proportionate heating demand Q_h (kWh/a) is defined with the help of the heat transfer coefficients (GRE, 2002).

$$Q_h = \text{ddn}^*\text{TFB}^*\text{CTR}^*A_{comp}^*k \qquad (4.3)$$

with

Q_h	Heating demand	—	kWh/a
ddn	Degree-day number, average value for Germany	—	84 Kd/a

---

[6]The k-value, or heat transfer coefficient, is the measured value of the heat flow which is transferred through an area of 1 m^2 at a temperature difference of 1 K. The units of measure are watts per square meter per temperature difference (W/m^{2}K). k-value = energy/(area × temperature difference × time). This value is used to measure the heat insulation efficiency of buildings.

TFB	Partial heating factor	—	0.9
CTR	Temperature reduction factor	—	1.0 (outer wall)
A_{comp}	Area of component	—	m^2
k	Heat transfer coefficient	—	$W/(m^2{*}K)$

Table 4.31 Parameters of insulation materials

	Unit	Without insulation	PS		VIP
Insulation thickness	mm	—	100	20	40
Heat conductivity of insulation material	W/(m*K)	—	0.035	0.004	0.004
k-value of outer wall	W/(m²*K)	1.504	0.285	0.177	0.094
Heating demand Q_h per 1 m² outer wall	kWh/a	113.7	21.5	13.4	7.1

Table 4.31 summarises the results. An insulation thickness of 20 mm is used for further calculation. The resulting energy saving by the use of VIPs in comparison to insulation with PS is calculated by multiplication of typical outer wall surfaces per building type by heating demand savings. Table 4.32 shows resulting energy savings per dwelling unit.

Table 4.32 Heat demand savings per housing unit

	Unit	Building with 1 hu	Building with 2 hu	Building with $\geq$ 3 hu
Heat demand saving per dwelling unit (hu) and building type	kWh/a	1,224.7	918.5	816.5

Quantitative impact on the energy demand in Germany is then estimated by multiplication of heat demand savings by the numbers of housing units insulated with VIPs.

The results of the scenario and sensitivity calculations for the possible impacts of VIPs are shown in Table 4.33. In 2030, between 23 PJ and up to 44 PJ could be saved in households by using innovative insulation materials based on nanotechnology.

Table 4.33   Energy savings due to insulation with VIPs

Energy savings (PJ)	2010	2015	2020	2025	2030
Nano −	0.0	0.0	1.7	9.3	23.3
Nano	0.0	0.2	3.1	14.5	35.1
Nano +	0.0	0.2	3.9	18.2	43.9

4.4.3.5 Polycarbonates for automotive glazing

One possibility to decrease fuel consumption of vehicles is to reduce vehicle weight. Lightweight vehicle design is gaining in importance and is considered to be a future trend in the automobile industry. In addition to the use of high-strength steel, light metals and fibre-reinforced synthetics automotive glazing could be employed. In the scope of the scenario calculations, the substitution of glass by polycarbonates for glazing is a matter of analysis. In order to determine the energy saving potential by the reduction of vehicle weight during operation, a correlation between vehicle weight and fuel consumption is needed. In literature, results that show great discrepancies can be found.

On the basis of a vehicle weight of 1,000 kg, according to Brink and Wee (1999), a decrease in fuel consumption of 7% per 100 kg of weight savings per 100 km is possible. This would mean a reduction in fuel consumption of about 0.5 l/100 km (with an average fuel consumption of 7.3 l/100 km). Klein (2002) gives a possible reduction of 0.3 up to 0.6 l/100 km per 100 kg weight saving, whereas Eberle (2000) considers a spectrum of 0.4 to 0.8 l. The weight of glass that could be substituted by polycarbonates is on average around 40 kg per vehicle (Möthrath *et al.*, 2007). The possible weight reduction by using polycarbonate as automotive glazing can range from 13 kg to 24 kg per vehicle. As a result, the possible reduction in fuel consumption varies between 0.7% and 1.2%, subject to the assumption that a decrease of vehicle weight by 100 kg results in a reduction in fuel consumption by 5%.

For market penetration it is assumed that in 2015, already 5% of automotive glazing will be made from polycarbonates. In 2030 this value will go up to 70%. The potential for polycarbonates is limited to the number of new cars. Table 4.34 shows the possible savings in fuel consumption. Starting from around 0.1 PJ in 2015, the energy savings increase to around 1.1 PJ till 2030 in the Nano scenario.

Table 4.34   Fuel savings due to the use of polycarbonates for automotive glazing

Energy savings (PJ)	2010	2015	2020	2025	2030
Nano −	0.0	0.0	0.1	0.3	0.6
Nano	0.0	0.1	0.2	0.5	1.1
Nano +	0.0	0.1	0.4	0.8	1.9

4.4.3.6 Nano-lacquers

On the basis of comprehensive data of a study conducted by Steinfeldt *et al.* (2004, 2007), it is possible to analyse and calculate the energetic impacts of manufacturing and application of nano-lacquers in the scenario and sensitivity analyses.

In doing so the manufacturing of the lacquer as well as its application is evaluated. On the basis of the different lacquer systems (1-/2-CCL, WCL, etc.), different amounts of lacquer to varnish one vehicle body are needed. In order to compare the use of a nano-lacquer with conventional lacquer systems, an average conventional lacquer is defined for the scenario analyses. This average lacquer is composed proportionately of the lacquer systems 1-/2-CCL, WCL, and PCL.

The energy demand for manufacturing and application is calculated in regard to the proportional energy demand of the different lacquer systems (see Table 4.35).

Table 4.35 shows that the energy demand of nano-lacquer in the fields of both manufacturing and application is lower than the average conventional lacquer system. Indeed the energy demand to manufacture 1 kg of nano-lacquers for application is higher than for conventional lacquer systems. However, due to a significant reduction in layer thickness and the possible reduction of processing steps, this effect can be offset already before the usage phase. For market penetration of nano-lacquers, it is assumed that it grew from 8% in 2015 and will grow up to 25% in 2030. Table 4.36 shows the possible impacts of nano-lacquers on energy consumption. Till 2030

Table 4.35   Energy demand for manufacturing and application of different lacquer systems

	Unit	1-CCL	2-CCL	WCL	PCL	Average conventional lacquer	Nano-lacquer
Energy demand manufacturing	(MJ/vehicle)	200	225	200	300	215	47
Energy demand application	(MJ/vehicle)	763	764	770	637	754	670
Total energy demand	(MJ/vehicle)	963	989	970	937	969	717
Amount of lacquer	(kg/vehicle)	2.3	2.3	3.2	2.4	2.7	0.3

up to 0.3 PJ could be saved during manufacturing and application of nano-lacquers.

Table 4.36   Energy savings in manufacturing and application of nano-lacquers

Energy savings (PJ)	2010	2015	2020	2025	2030
Nano −	0.0	0.1	0.1	0.1	0.2
Nano	0.0	0.1	0.1	0.2	0.2
Nano +	0.0	0.1	0.2	0.2	0.3

The possible application of nano-lacquers has been on the agenda in automotives for a long time. According to Steinfeldt *et al.* (2004, 2007), there seems to be a saving potential of using nano-lacquers in comparison to conventional lacquers not just in manufacturing and application but also during operation of vehicles. The thinner lacquer layers of nano-lacquers in comparison to conventional lacquer systems (e.g., 1-/2-CCL, WCL, etc.) results in a weight reduction of the vehicle.

On the basis of the correlation that was used to determine the impacts of polycarbonates (see Section 4.4.3.5), the possible reduction in fuel consumption by the use of nano-lacquers is derived. By the use of nano-lacquers the weight of a car body can be reduced by around 2.4 kg.

This results in a possible reduction in fuel consumption of around 0.12%. It is assumed that in 2015 around 5% of all new cars will be varnished with nano-lacquers and in 2030 around one-fourth of all new vehicles. The results of the scenario calculations are shown in Table 4.37. The use of nano-lacquers has only a slight impact on energy consumption in automotives in comparison to manufacturing and application. According to the calculations up to 0.07 PJ could be saved until 2030.

Table 4.37 Fuel savings due to the use of nano-lacquers in automotives

Energy savings (PJ)	2010	2015	2020	2025	2030
Nano –	0.00	0.01	0.02	0.03	0.04
Nano	0.00	0.01	0.03	0.04	0.05
Nano +	0.00	0.01	0.03	0.05	0.07

4.4.3.7 Nanocatalysts for styrene manufacturing

Almost 80% of classical chemical processes on an industrial scale are based on the use of catalysts. Analyses show that 25% of the gross value added in Western industrialised countries is based on catalysts (DFG, 2006). By now, catalysts can accelerate chemical reactions which subsequently increase the rate of yield and result in fewer by-products (Heubach *et al.*, 2005). Nanotechnology gives a starting point to further enhance catalytic reaction due to the fact that on the nanoscale, the ratio between surface and volume of the catalyst increases.

The production of styrene is one of the 10 most important petrochemical processes because it is one of the most important basic chemical products in the chemical industry. The annual German production in 2010 was 0.954 million tones (DESTATIS, 2011). Steinfeldt *et al.* (2004, 2007) reported energetic impact of nano-catalysts on styrene manufacturing. For the classical synthesis a catalyst based on iron oxide is used (MPG, 2002). With the use of a nanostructured multi-walled carbon nanotube, this endothermic process can be transferred into an exothermic process.

The production amount can be increased, and operating temperature can be lowered to about 200°C. With the help of a nanoscale catalyst, the overall specific energy consumption to manufacture 1 kg styrene can be lowered from 6.36 MJ to 3.25 MJ (Steinfeldt *et al.*, 2004, 2007). On the basis of German styrene production, the variation of growth rate (3.8% to 6.3%), the variation of future enhancements in energy efficiency in the production processes (1.5% to 2.5%) for conventional styrene manufacturing, the market share of nanotubes as catalysts, and the possible reduction in energy consumption due to the use of the nanocatalyst and the possible impact by its use are analysed and calculated. It is assumed that in 2015, already 5% of all manufacturing processes will be based on the nanocatalyst, and in 2030 half of the styrene production will be manufactured with the help of a catalyst based on nanotubes. Table 4.38 shows the possible energy savings in styrene manufacturing through the use of nanocatalysts. It seems conceivable that till 2030 around 3.7 PJ could be saved.

Table 4.38   Energy savings in styrene manufacturing

Energy savings (PJ)	2010	2015	2020	2025	2030
Nano −	0.0	0.2	0.4	1.0	2.2
Nano	0.0	0.2	0.6	1.6	3.7
Nano +	0.0	0.3	0.8	2.4	5.7

4.4.3.8   Nanoparticles in synthetic production

As described in Section 4.3.3.8, the use of nanoparticles can increase flowability of PBT considerably, and therefore the production processes of injection moulding can be optimised. In 2009 around 0.75 million tons of PBT were produced worldwide. The yearly growth rate of PBT is around 5%, and in Germany around 10% of the worldwide production is processed. About 92% of the PBT is used for injection moulding. The energy demand of the electrical process to produce 1 kg PBT in injection moulding processes is around 3 kWh (BASF, 2004; Eibeck and Anderlik, 2010). By the addition of nanoparticles, melt viscosity of PBT can be reduced notably.

Therefore it is possible to produce more thin-walled components, cycle times can be reduced, and injection pressures can be lowered. All in all, a reduction in energy demand of around 20% can be assumed (BASF, 2005; Eipper and Völkel, 2006; Weiß and Eipper, 2004). In the scope of the scenario and sensitivity analyses, the possible energy reduction is varied in a range of 11% to 20%. It is assumed that in 2015 market penetration is already 30% and in 2030 nanoparticles will be used in the total amount of PBT processed in injection moulding.

On the basis of the amount of PBT processed in injection moulding, future market penetration, as well as the possible energy reduction, the impact on energy demand for the manufacturing of PBT in injection moulding is calculated for synthetics production in Germany.

Table 4.39 shows the impacts of nanoparticles on the energy demand in synthetic production till 2030. It shows that around 0.4 PJ could be saved by the use of nanoparticles.

Table 4.39   Energy savings in synthetic production

Energy savings (PJ)	2010	2015	2020	2025	2030
Nano −	0.0	0.0	0.1	0.1	0.2
Nano	0.0	0.1	0.2	0.3	0.4
Nano +	0.0	0.1	0.4	0.6	0.8

4.4.3.9 Nanoparticles in tyre compounds

According to Velte (2006) and Unseld (2006), a reduction in rolling friction of 15% to 25% by the use of carbon nanotubes as tyre fillers in tyre compounds seems possible.

This would result in fuel consumption savings of around 4%, depending on the findings of König (2004) that a reduction in rolling friction of around 30% leads to a saving in fuel consumption of 5%.

In the scope of the scenario and sensitivity analyses, a possible fuel consumption reduction varies between 0.8% and 4%. Furthermore the technical potential is determined to be 100% for all tyres. An estimation of a possible market penetration for such new tyre fillers and the associated savings in fuel consumption is at present still very difficult as most manufacturing methods to generate nanostructures in tyre compounds still have to be further developed (Velte, 2006). However, in the scenario and sensitivity calculations, it is positively assumed that in 2015, already 15% of all tyres will use fillers based on nanotechnology. This percentage will grow till 2030 to about 70%. According to Table 4.40 it is expected that by the use of nanoparticles in tyre compounds, up to 35 PJ could be saved in 2030.

Table 4.40   Fuel savings due to nanoparticles in tyre compounds

Energy savings (PJ)	2010	2015	2020	2025	2030
Nano –	0.0	1.1	3.2	6.1	9.5
Nano	0.0	7.4	19.8	29.6	34.5
Nano +	0.0	11.4	30.9	53.4	67.3

4.4.3.10 Nano-based coatings to reduce friction

Despite ongoing reduction in production tolerances and further developments in the field of lubricants, 15% of fuel consumption of modern automotive engines still relies on engine friction (HL, 2006).

Analyses based on the New European Driving Cycle (NEDC) show that a reduction in fuel consumption of vehicles of about 11% is possible (Thesenvitz *et al.*, 2007). In the scope of the scenario and sensitivity calculations, the possible fuel consumption reduction varies between 6% and 11%. The technical potential to use nano-based coatings to reduce friction in the engine, diesel-driven and petrol driven vehicles is assumed to be 100%.

In fuel cell vehicles such coatings can only be applied to the power train, and in that case the possible potential cannot be exploited. In other words, the technical potential for fuel cell vehicles is determined to be 25%.

In 2015 about 3% of the total vehicle fleet will apply nano-based coatings, and this value will grow up to around 9% in 2030. The results of the scenario and sensitivity calculations for the use of nano-based coatings to reduce friction are summarised in Table 4.41. In 2030 between 8.4 PJ and up to 24.6 PJ could be saved in the traffic sector by the use of nano-based coatings that reduce friction in engines and drive trains.

Table 4.41   Fuel savings due to nano-based coatings to reduce friction

Energy savings (PJ)	2010	2015	2020	2025	2030
Nano −	0.0	2.8	4.6	6.5	8.4
Nano	0.0	4.9	8.2	11.5	14.8
Nano +	0.0	8.2	13.7	19.2	24.6

Bibliography

Alber, S. (2006). Personal message, 04.07.2006.

Bachmann, G., Grimm, V., Hoffknecht, A., Luther, W., Ploetz, C., Reuscher, G., Teichert, O., and Zweck, A. (2007). Nanotechnologie für den Umweltschutz, Zukünftige Technologien Consulting, VDI Technologiezentrum, Düsseldorf.

BASF (2004). Daten und Fakten 2004: BASF Gruppe, Stand: April 2004, BASF AG, Ludwigshafen.

BASF. (2005). Label Eco-Efficiency Analysis Ultradur® High Speed, BASF Aktiengesellschaft, Ludwigshafen.

BASt. (2006). Verkehrs- und Unfalldaten: Kurzzusammenstellung der Entwicklung in der Bundesrepublik Deutschland, Bundesanstalt für Straßenwesen.

BDZ. (2011). Zahlen und Daten, Bundesverband der Deutschen Zementindustrie e.V., Berlin.

Bernreuter, J. (2002). Ein riesiges Potenzial - Killerargumente wiederlegen: 'Photovoltaik bleibt in Deutschland unbedeutend', PHOTON das Solarstrom-Magazin, September.

Birnbaum, U., Bongartz, R., Linssen, J., Markewitz, P., Vögele, S. (2010). Gemeinschaft STE Research Report -Energietechnologien 2050 – Schwerpunkte für Forschung und Entwicklung, Fossil basierte Kraftwerkstechnologien, Wärmetransport, Brennstoffzellen, Jülich: Forschungszentrum Jülich, Institut für Energieforschung, Systemforschung und Technologische Entwicklung (IEF-STE).

BMWi. (2011). Zahlen und Fakten: Energiedaten: Nationale und Internationale Entwicklung, Stand: 07.12.2011, Bundesministerium für Wirtschaft und Technologie, Berlin.

Brink, R. V. d., and Wee, B. V. (1999). Passenger car fuel consumption in the recent past: Why has passenger car fuel consumption no longer shown a decrease since 1990?, Paper prepared for the workshop

Indicators of Transportation Activity, Energy and CO_2 emissions, May 9–11, Stockholm.

Daimler. (2009). Neuer Mercedes-Benz-Hybridbus im Praxistest, Daimler Communications, Mercedes-Benz - Eine Marke der Daimler AG, Stuttgart.

Daimler. (2010a). Forschungsfahrzeug F600 HYGENIUS, Daimler AG, Stuttgart.

Daimler. (2010b). Drei Mercedes-Benz Citaro G BlueTec-Hybrid-Busse jetzt bei Stuttgarter Straßenbahnen im Einsatz, Daimler Communications, Mercedes-Benz - Eine Marke der Daimler AG, Stuttgart.

Daimler. (2011). Hybridbus für Landkreis München, Daimler Communications, Mercedes-Benz - Eine Marke der Daimler AG, Stuttgart, München.

DESTATIS. (2011). GENESIS-Online, Datenbank, Statistisches Bundesamt Deutschland, Wiesbaden.

DFG. (2006). Jahresbericht 2006 - 'Heiratsvermittler' der Chemie: die Katalyse, DFG (Deutsche Forschungsgemeinschaft).

Eberle, R. (2000). Methodik zur ganzheitlichen Bilanzierung im Automobilbau, Institut für Straßen- und Schienenverkehr, TU Berlin, Berlin.

Edel, M. (2006). Personal message, 03.08.2006.

Edel, M. (2003). Brennstoffzellen für die Hausenergieversorgung -Technologie, Entwicklungsstand, Einsatzmöglichkeiten, EnBW AG Energie-Vertriebsgesellschaft mbH, Stuttgart.

Eibeck, P. and Anderlik, R. (2010). Polybutylenterephthalat (PBT), Kunststoffe, Spezial K 2010, Carl Hanser Verlag, München.

Eipper, A. and Völkel, M. (2006). Filigrane Bauteile wirtschaftlich herstellen, Kunststoffe - Nanotechnologie (11/2006).

ETAP. (2007). Nanotechnology brings new Solutions for Carbon Capture, Environmental Technologies Action Plan (ETAP).

Fischle, H. J. (2005). Superkondensatoren, made by WIMA, WIMA Spezialvertrieb elektronischer Bauelemente GmbH & Co.KG, Berlin.

Fröhlich, K. (2007). Projekt – Anwendungspotential der thermoelektrischen Stromerzeugung im Hochtemperaturbereich, Jahresbericht 2007; ETH Zürich, Institut für elektrische Energieübertragung und Hochspannungstechnologie, Zürich.

Geitmann, S. (2005). DaimlerChrysler plant Marktreife von Brennstoffzellenautos bis 2015, Energieportal24.de.

GRE. (2002). Energieeinsparung im Gebäudebestand: Bauliche und anlagentechnische Lösungen, Gesellschaft für Rationelle Energieverwendung e.V., Berlin.

Gummert, G., and Suttor, W. (2006). Stationäre Brennstoffzellen: Technik und Markt, Müller (C.F.), Heidelberg.

Hamm, M. (2006). LED im Audi R8 - was ist die technische Herausforderung? Automotive Lightning, Reutlingen.

Heubach, D., Beucker, S., and Lang-Koetz, C. (2005). Einsatz von Nanotechnologie in der hessischen Umwelttechnologie: Innovationspotenziale für Unternehmen, HA Hessen Agentur GmbH, Wiesbaden.

HL. (2006). Dokumentation Nano - Hier ist die Zukunft - Hessen im Dialog, Hessische Landesregierung.

Hofmann, T. (2006). Nanobeschichtungen für Architektur- und Solargläser, Vortrag vom 24.01.2006, VDI Technologiezentrum, Nanotecture 2006 – Anwendungen der Nanotechnologie in Architektur und Bauwesen, Düsseldorf.

IEA. (2005). Energy Technology Analysis: Prospects for Hydrogen and Fuel Cells, International Energy Agency, Paris.

Jopp, K. (2006). Nanotechnologie — Aufbruch ins Reich der Zwerge, Betriebswirtschaftlicher Verlag Dr. Th. Gabler GWV Fachverlage GmbH, Wiesbaden.

Jüstel, T. (2006). Neuartige Leuchtstoffe für Hochleistungs-LEDs, FB Chemieingenieurwesen, FH Münster, Abt. Steinfurt.

Kaltschmitt, M., Streicher, W., and Wiese, A. (2006). Erneuerbare Energien - Systemtechnik, Wirtschaftlichkeit, Umweltaspekte, Springer-Verlag, Berlin, Heidelberg, New York.

Klein, B. (2002). Leichtbau im Automobilbau, Universität Kassel, Fachbereich Maschinenbau, Fachgebiet Leichtbau-Konstruktion, Kassel.

König, U. (2004). Beiträge der Nanotechnologie für umweltfreundliche Automobile, Symposium Nano meets Umwelttechnik, Fraunhofer IAO Stuttgart, Vortrag vom 02.07.2004.

Kreschl, S., Hipp, E., and Lexen, G. (2005). Effizienter Hybridantrieb mit Ultracaps für Stadtbusse, Vortrag beim 14. Aachener Kolloquium Fahrzeug- und Motorentechnik, Aachen.

MAN. (2005). Optimierter MAN-Ultracap-Bus im Linieneinsatz bei der VAG in Nürnberg: Umweltfreundliches Hybridkonzept mit Nutzung der Bremsenergie, MAN Nutzfahrzeuge AG.

Möthrath, S., Köppchen, and Krause. (2007). Polycarbonatanwendungen im Automobilbau, Bayer AG.

MPG. (2002). Nanozwiebeln bringen Styrol-Synthese auf Trab, Max-Planck Gesellschaft.

Nitsch, J., Pregger, T., Scholz, Y., Naegler, T., Sterner, M., Gerhardt, N., Oehsen, A., Pape, C., Saint-Drenan, Y.-M., Wenzel, B. (2010). Langfristszenarien und Strategien für den Ausbau der erneuerbaren Energien in Deutschland bei Berücksichtigung der Entwicklung in Europa und global, Leitstudie 2010, Deutsches Zentrum für Luft- und Raumfahrt (DLR), Institut für technische Thermodynamik Abt. Systemanalyse und Technikbewertung, Fraunhofer Institut für Windenergie und Energiesystemtechnik (IWES), Ingenieurbüro für neue Energien (IFNE), Bundesministerium für Umwelt, Naturschutz und Reaktorsicherheit (BMU), Berlin.

OLED100. (2009). OLED100.eu - Project Report Three aesthetical perception case studies, Organic LED Lighting in European Dimensions funded under the IST priority.

OLED100. (2011). Organic LED lighting in European dimensions, Homepage: http://www.oled100.eu/homepage.asp.

OLLA. (2009). OLLA Project Report - Final Activity Report, High brightness OLEDs for ICT & Next Generation Lighting Applications funded under the IST priority.

Paschen, H., Coenen, C., Fleischer, T., Grünwald, R., Oertel, D., and Revermann, C. (2004). Nanotechnologie - Forschung, Entwicklung, Anwendung, Springer Verlag, Berlin, Heidelberg, New York.

Pester, W. and Trechow, P. (2009). Bordstrom direkt aus Autoabgas zapfen, VDI-Nachrichten, Nr. 44(30 October 2009).

Quaschning, V. (2008). Erneuerbare Energien und Klimaschutz: Hintergründe- Techniken-Anlagenplanung- Wirtschaftlichkeit, Carl Hanser Verlag, München.

Schiermeister, N., Schwenkschuster, L., Decker, D., and Eichhorn, K. (2003). Leuchtdioden-Systeme im Scheinwerfer, ATZ (Automobiltechnische Zeitung), 105, 2003, 9, Sonderdruck, S. 1–11.

Steinfeldt, M., Gleich, A. V., Petschow, U., Haum, R., Chudoba, T., and Haubold, S. (2004). Nachhaltigkeitseffekte durch Herstellung und Anwendung nanotechnologischer Produkte, Institut für ökologische Wirtschaftsforschung (IÖW) gGmbH.

Steinfeldt, M., Gleich, A. V., Petschow, U., and Haum, R. (2007). Nanotechnologies, Hazards and Ressource Efficiency - A Three-Tiered Approach to Assessing the Implications of Nanotechnology and Influencing its Development, Berlin, Heidelberg: Springer Verlag.

Stobbe, L., Nissen, N. F., Proske, M., Middendorf, A., Schlomann, B., Friedewald, M., Georgieff, P., and Leimbach, T. (2009). Abschätzung des Energiebedarfs der weiteren Entwicklung der Informationsgesellschaft, Abschlussbericht an das Bundesministerium für Wirtschaft und Technologie, Karlsruhe.

Strategierat. (2006). Nationaler Entwicklungsplan Version 1.1 zum, Innovationsprogramm Wasserstoff- und Brennstoffzellentechnologie, Strategierat Wasserstoff Brennstoffzellen.

Thesenvitz, M., Mohn, B., and Bohl, B. (2007). Prüfbericht: Nanoprotect, Fachhochschule Frankfurt am Main, Labor für Kraftfahrzeugtechnik, Frankfurt am Main.

UBA_CH. (2004). Handbuch für Emissionsfaktoren, Version 2.1, Umweltbundesamt Schweiz, Bern.

UH. (2008). Umwandlung von Wärme in Elektrische Energie, Pressedienst Universität Hamburg.

Unseld, K. (2006). Personal message, 10.07.2006.

Velte, G. (2006). Personal message, 03.07.2006.

Viebahn, P., Esken, A., Höller, S., Luhmann, J., Pietzner, K., Vallentin, D., Dietrich, L., and Nitsch, J. (2010). RECCS plus Regenerative Energien (RE) im Vergleich mit CO_2-Abtrennung und -Ablagerung (CCS) - Update und Erweiterung der RECCS-Studie, Forschungsvorhaben gefördert vom Bundesministerium für Umwelt, Naturschutz und Reaktorsicherheit, Wuppertal.

Weiß, C., and Eipper, A. (2004). Neues leicht fließendes PBT durch Nanotechnologie, *Plastverarbeiter*, 55. Jahrg.(Nr. 10), S. 178–180.

Wilk, A. (2006). Personal message, 18.07.2006.

4.5 Comprehensive Subsumption of Nanotechnology in the Energy Sector

Jochen Lambauer, Dr. Ulrich Fahl, and
Prof. Dr. Alfred Voß
Institut für Energiewirtschaft
und Rationelle Energieanwendung (IER),
Universität Stuttgart, Germany

In Section 4.3.4 the theoretical potential of nanotechnological applications was evaluated. In the preceding paragraphs the approach and the results of the scenario and sensitivity calculations

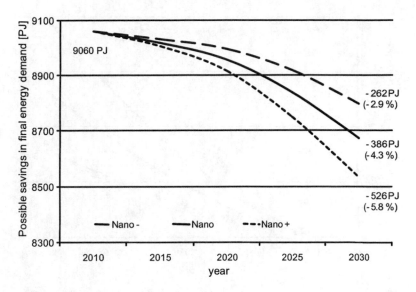

Figure 4.12 Development of possible savings in final energy demand from 2010 to 2030.

have been described in detail. In order to get a broad overview about the possible impacts of nanotechnology on the energy sector in Germany, the theoretical potentials and the findings from the scenario and sensitivity calculations are used as a basis for comprehensive evaluation and subsumption.

Figure 4.12 shows the development of possible energy savings from 2005 till 2030. Due to uncertainties in regard to future development of nanotechnological application, market penetration, and future energy demand, different scenario and sensitivity calculations are conducted. The Nano scenario summarises the expected development. Within the sensitivitiy analyses input parameters such as market penetration or prevalence rate are varied by +25% (Nano +) and −25% (Nano −).

Till 2030 the possible potentials for savings in final energy demand range from 262 PJ to 526 PJ. Compared with the final energy demand in Germany in 2010 (9,060 PJ), this could be a reduction of 3% to 6% in 2030. The application of nano-based membranes for CCS could result in energy savings ranging from 195 PJ to 325 PJ. Figure 4.13 shows, similar to Figure 4.12, the development of

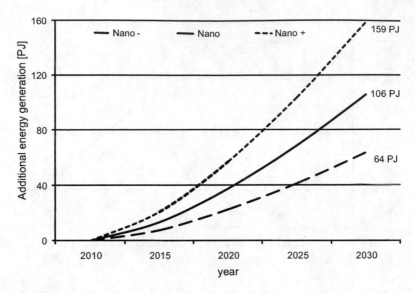

Figure 4.13 Development of possible additional energy generation from 2010 to 2030.

additional energy generation due to nanotechnological applications till 2030 for the scenario and sensitivity calculations.

Additional energy generation ranges from 64 PJ up to about 159 PJ in 2030 by the use of antireflective coatings, new photovoltaic technologies, and the application of TEGs in power plants.

For both developments it can be concluded that the impacts of the analysed nanotechnological applications on energy consumption as well as on energy generation require a long time until significant impacts are visible. In 2020 the possible reductions correspond to 1.1% of the final energy consumption in 2010 in the Nano scenario.

In conclusion, the evaluated applications together present a significant potential for energy savings in Germany till 2030. But it is obvious that it will take quite a long time till the positive effects are visible.

The development in potentials for additional energy generation is closely similar to the possible potential for energy savings. But it takes even longer until effects of nanotechnological applications on energy generation and conversion become significant.

Table 4.42 shows the results of the scenario calculations for the expected development in the Nano scenario till 2030.

The results of the scenario calculations show that the analysed 26 nanotechnological applications have the potential to reduce energy consumption by about 647 PJ till 2030. On the basis of the evaluated antireflective coatings, new photovoltaic technologies, and the application of TEGs in power plants, an additional 106 PJ of electricity could be generated. The main impact on additional energy generation is attributable to the use of thin-film solar cells. In sum they could generate almost up to 50 PJ in 2030. In regard to energy savings, the major potential shows the application of nano-based membranes for CCS (260 PJ) followed by the replacement of conventional lighting technologies with LEDs and OLEDs (153 PJ) and the use of stationary fuel cells in households (so-called fuel cell heating units).

In the traffic sector the application of nanoparticles in tyre compounds (35 PJ) or the use of the fuel additive ceroxide (35 PJ) shows high potential. In case of ceroxide, we have to take into account the possible effects on the environment by burning ceroxide. Up to now still insufficient information about the possible effects on the environment is available. The application of OLED displays (10 PJ) or the use of nanoparticles in synthetic production (0.4 PJ) shows, in contrast, minor potential for energy savings. In the first case this is due to the small number of OLED displays that will already be in use in 2030. The comparatively small potential for the use of nanoparticles in synthetic production results from the small reference value, the energy demand for synthetic production.

Table 4.43 compares the expected potential (Nano scenario) with the theoretical potential (see Section 4.3.4) till 2030. The comparison clearly shows that in the traffic sector and in industry, there are still high unused potentials for energy savings in future. In the traffic sector only around 11% of the theoretical potential will be captured till 2030. A similar picture in regard to energy generation happens in the generation and conversion sector. Till 2030 around 26% of the theoretical potential seems possible.

Table 4.42 Change from 2010 to 2030 along the energy value chain (Nano scenario)

Application		Nano scenario till 2030 and share in regard to theoretical potential					
		Energy generation		Energy conversion, storage, and distribution		Energy usage	
		(PJ)	(%)	(PJ)	(%)	(PJ)	(%)
Antireflective coatings	Solar thermal	2.8	95.0				
	CSP	0.8	74.3				
	Photovoltaics	0.7	95.0				
New photovoltaic technologies	Thin-film solar cells	48.8	25.0				
	Dye solar cells	20.3	25.0				
	Organic solar cells	6.7	22.6				
	Staple solar cells	6.0	25.0				
	Quantum dot solar cells	5.6	25.0				
Fuel cell heating units				81.5	49.5		
Fuel cell vehicles				10.2	1.6		
Ceroxide				34.8	55.5		
Nano-based membranes for CCS				260.2	63.7		
TEGs	Automotives					1.1	50.7
	Power plants	14.2	30.0				
Supercapacitors in hybrid buses						0.8	27.7
LEDs for multiple-coach lighting						115.2	85.5
LEDs in automotive lighting						4.6	92.4
OLEDs and LEDs for multiple-coach lighting						153.1	74.1
OLED displays						10.6	32.0
UHPC						0.4	2.5
Insulation with VIPs						35.1	29.3
Polycarbonates for automotive glazing						1.1	70.0
Nano-lacquers	Manufacturing and application					0.2	25.0
	Operation					0.0	25.0

(Contd.)

Table 4.42   (*Contd.*)

Application	Nano scenario till 2030 and share in regard to theoretical potential					
	Energy generation		Energy conversion, storage, and distribution		Energy usage	
	(PJ)	(%)	(PJ)	(%)	(PJ)	(%)
Styrene manufacturing					3.7	50.0
Nanoparticles in synthetic production					0.4	100.0
Nanoparticles in tyre compounds					34.5	70.0
Nano-based coatings to reduce friction					14.8	9.1

Table 4.43. Results of the Nano scenario till 2030 and theoretical potential for different sectors

	Generation/ conversion	Industry	Tertiary sector	Households	Traffic
Development till 2030 (Nano scenario)					
Energy savings (PJ)	260.2	26.2	108.7	150.0	101.8
Additional energy generation (PJ)	105.8				
Theoretical potential					
Energy savings (PJ)	408.6	55.6	148.1	346.8	916.5
Additional energy generation (PJ)	404.0				

This result shows that in regard to energy consumption, nanotechnology has the potential for energy savings already till 2030.

In regard to energy generation and conversion, nanotechnology is starting to add value already till 2030, but there seems to be further great potential for additional energy generation due to nanotechnological application after 2030.

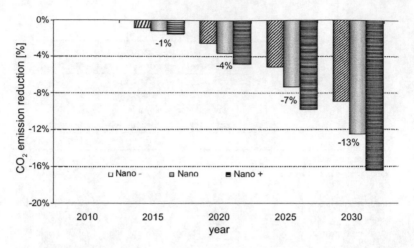

Figure 4.14 Reduction of CO_2 emissions due to energy savings and additional energy generation.

Next to enhancements in energy efficiency and energy savings, reduction of CO_2 emissions is another key challenge in the future. In this regard Figure 4.14 shows the possible development of reduction of CO_2 emissions through the evaluated nanotechnological applications in regard to energy savings and energy generation (antireflective coatings, new photovoltaic technologies, and TEGs in power plants). Compared with CO_2 emissions in Germany in 2010 (777 million tons), a reduction of almost 97 million tons CO_2 (12.5%) in 2030 seems conceivable. The evaluations from Cientifica (2007a,b) and Luther (2008) show similar results about the possible impacts and effects of nanotechnology on the energy sector. According to Cientifica (2007a,b) possible worldwide impacts of applications of nanotechnology-based tools resulted in a reduction in CO_2 emissions of around 0.2 million tons in 2010. These reductions were mainly attributable to the improvements in the building sector resulting from new insulation technologies as well as in the traffic sector from improvements in combustion technologies. Both Cientifica (2007a,b) and Luther (2008) came to the conclusion that nanotechnology has great potential for energy savings and reduction of CO_2 emission but not in the short term.

Table 4.44.   Comparison between the Nano scenario and projections according to Fahl *et al.* (2010)

Sector	Energy savings till 2030 (Nano scenario)		Reference value 2030*
	(PJ)	(%)	(PJ)
Additional energy generation	105.8	4.7	2236
Industry	26.2	1.2	2130
Tertiary sector	108.7	8.8	1241
Households	150.0	7.6	1982
Traffic	101.8	4.2	2442

*Please refer to Fahl *et al.* (2010).

On the basis of the study 'Projecting Energy Market Trends until 2030-Energy Outlook 2009', (Fahl *et al.*, 2010), commissioned by the German Federal Ministry of Economics and Technology, Table 4.44, compares the results of the Nano scenario with the projected development of energy demand in Germany in 2030 (reference case). The absolute values are the results from the Nano scenario with the impact of nanotechnological applications till 2030. The analysed nanotechnological application could cover almost 5% of electricity generation in 2030. In regard to energy savings nanotechnology could reduce energy consumption, for example, in households in 2030, by about 7.6%. Savings of more than 4% could be possible in the traffic and tertiary sectors.

Nanotechnology can offer undeniably great opportunities for an eco-efficient economy and a sustainable energy system. The analyses show that the potential benefits of nanotechnology in terms of emission reduction, sustainable energy generation, and new energy and resource-efficient production systems and products are wide ranging and immense. As a cross-cutting key enabling technology, there is no limit on the potential application of nanotechnology. On the other hand, the feasibility of numerous applications in industrial-scale production has still to be further examined.

In addition to the possibilities and potentials of energy savings and emission reduction brought by nanotechnology, the impact of nanotechnological materials on the environment and humans is as well decisive and has to be taken into consideration before new technological applications can be substantiated. As history shows that new technologies can have negative impacts on the environment, human health, society, or the economy that were not intended (see Section 2.4), safety should be an integral part of all nanotechnology-based innovations and products in the planning. Therefore dedicated research on potential risks has to be linked to existing and future applications of nanotechnology but correlated to benefits.

Next to potential energy savings, the energy demand of manu-facturing and production of nanotechnological applications has to be taken into account. The total life cycle aspects of nanomaterials and general resource efficiency have to be a matter of analysis when judging the eco-friendliness of nanomaterials. Possible energy efficiency potentials of nanotechnology in production processes are described in detail in Section 3.4.3. During the last years, more and more LCAs on nanotechnology-based products, and applications have been conducted (see Section 4.2). These LCAs show that nanotechnological applications as a whole cannot be associated with a large potential for environmental relief. Nevertheless, potentials of selected applications for significant environmental relief can be ascertained.

This initial potential analysis shows that it takes a long time before already close-to-market applications result in significant reduction in energy consumption and CO_2 emissions. In regard to the challenges concerning the use of energy, it is necessary to work intensely and to prompt market introduction of innovative energy technologies in order to solve these given challenges. Nanotechnology is not going to change the energy sector radically, but through the development of innovative products, processes, and applications, it is going to contribute to cost reduction, to the rational use of energy, and to energy efficiency improvements.

Bibliography

Cientifica. (2007a). Nanotech and Cleantech Reducing Carbon Emissions Today, Cientifica, London.

Cientifica. (2007b). Nanotech: Cleantech Quantifying the Effect of Nanotech-nologies on CO_2 Emissions, Cientifica, London.

Fahl, U., Blesl, M., Voß, A., Achten, P., Bruchof, D., Götz, B., Hundt, M., Kempe, S., Kober, T., Kuder, R., Küster, R., Lambauer, J., Ohl, M., Remme, U., Sun, N., Wille, V., Wissel, S., Ellersdorfer, I., Kesicki, F., Frondel, M., Grösche, P., Peistrup, M., Ritter, N., Vance, C., Zimmermann, T., Löschel, A., Bühler, G., Hoffmann, T., Mennel, T., and Wölfing, N. (2010). Die Entwicklung der Energiemärkte bis 2030 : Energieprognose 2009, Institut für Energiewirtschaft und Rationelle Energieanwendung (IER), Stuttgart.

Luther, W. (2008). Einsatz von Nanotechnologien im Energiesektor, HA Hessen Agentur GmbH, Wiesbaden.

Index

Colour Insert

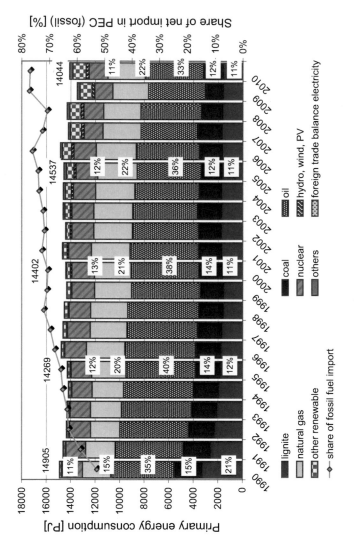

Figure 1.1

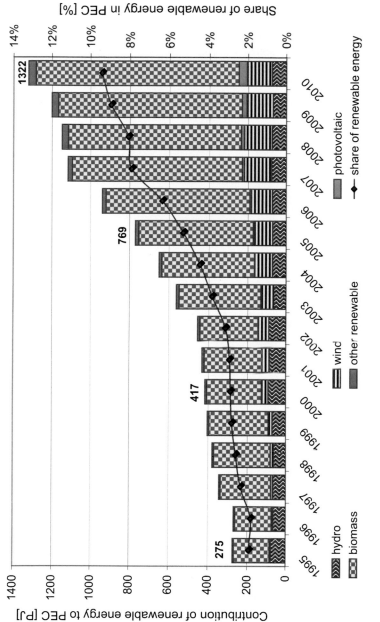

Figure 1.2

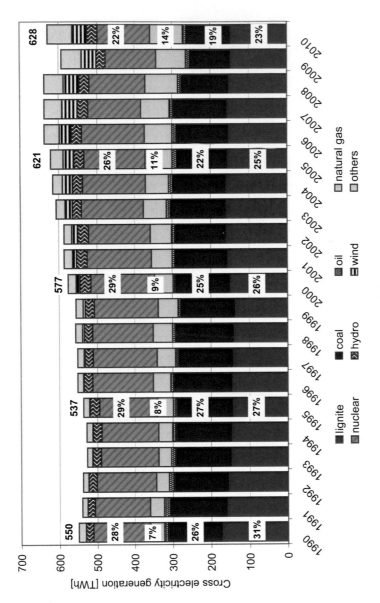

Figure 1.3

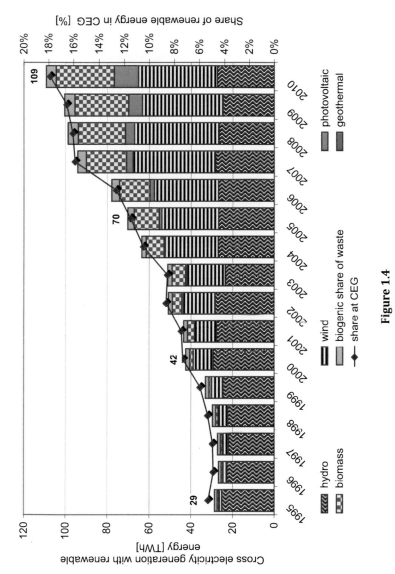

Figure 1.4

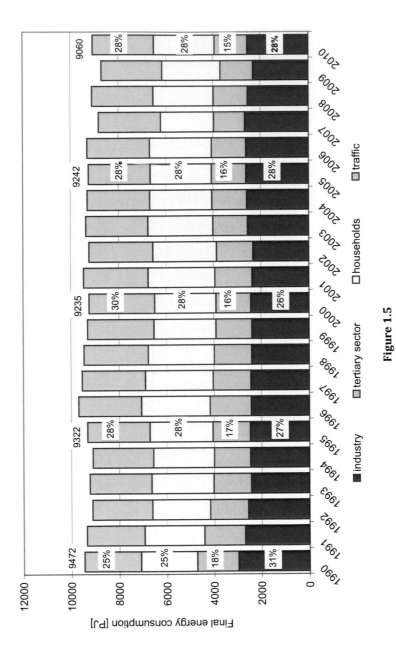

Figure 1.5

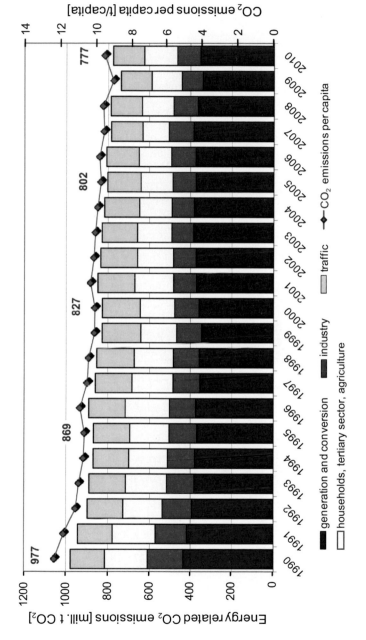

Figure 1.6

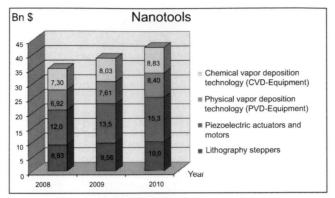

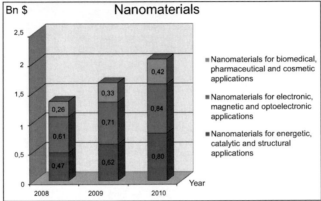

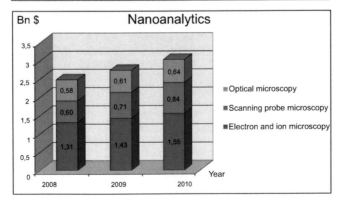

Figure 2.4

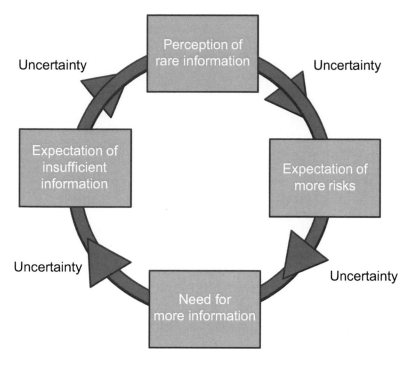

Figure 2.5

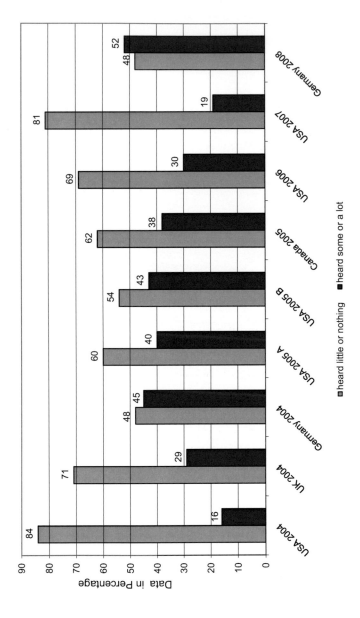

Figure 2.6

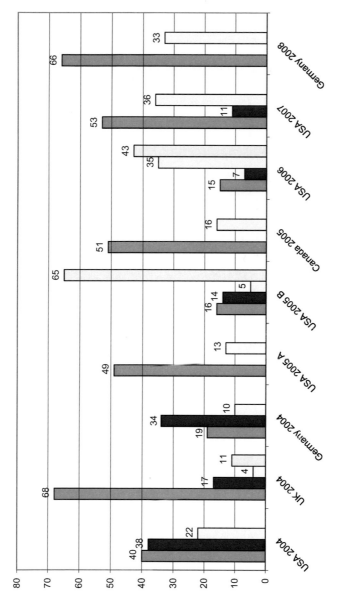

Figure 2.7

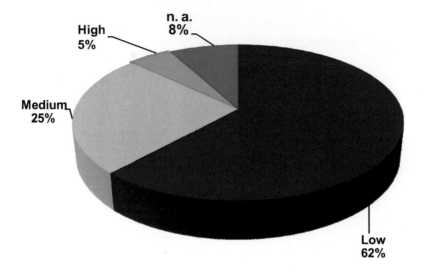

Figure 2.8

Would you try a nanoproduct? N=100

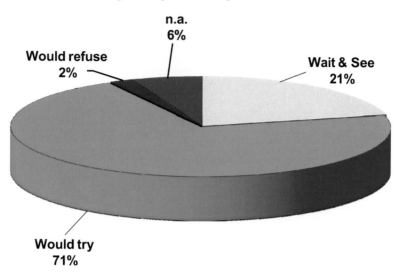

n.a.
6%

Would refuse
2%

Wait & See
21%

Would try
71%

Attitudes towards nanotechnologies? N=100

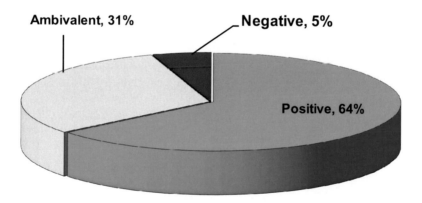

Ambivalent, 31%

Negative, 5%

Positive, 64%

Figure 2.10

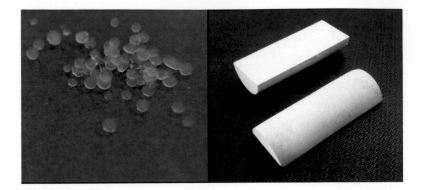

Figure 3.2

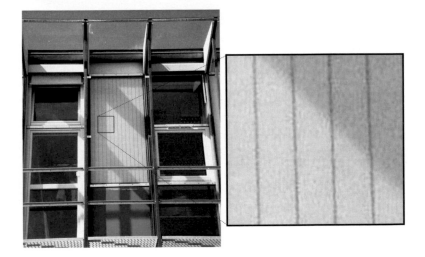

Figure 3.4

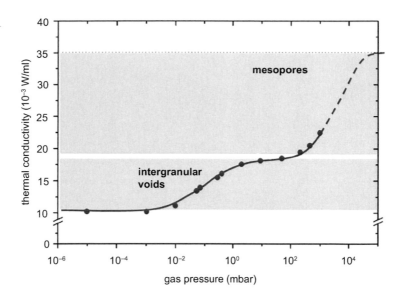

Figure 3.5

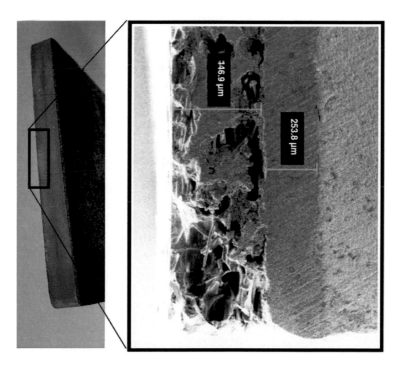

Figure 3.8

Figure 3.9

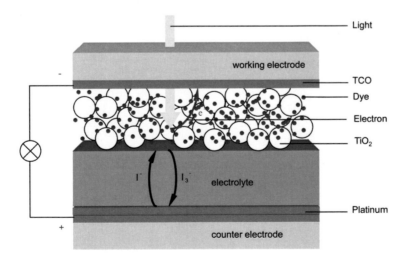

Figure 3.13

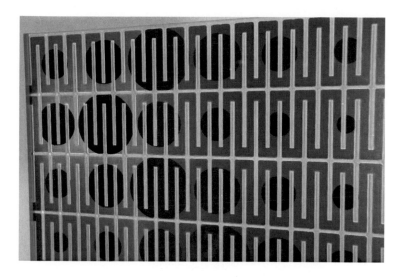

Figure 3.14

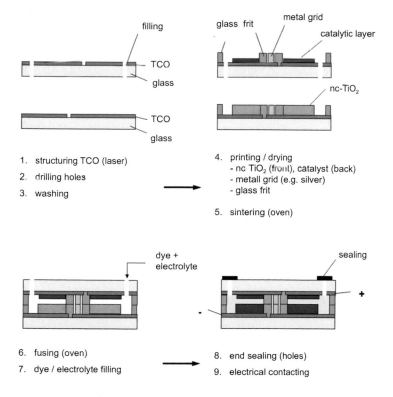

1. structuring TCO (laser)
2. drilling holes
3. washing

4. printing / drying
   - nc TiO₂ (front), catalyst (back)
   - metall grid (e.g. silver)
   - glass frit

5. sintering (oven)

6. fusing (oven)
7. dye / electrolyte filling

8. end sealing (holes)
9. electrical contacting

Figure 3.16

Figure 3.17

Figure 3.18

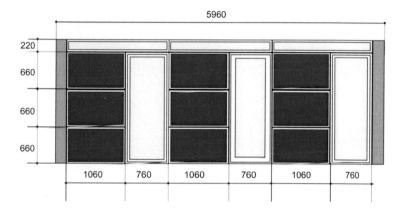

Figure 3.19

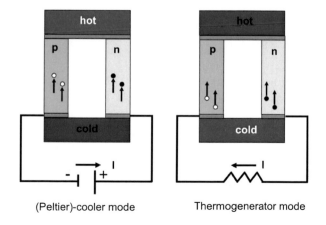

(Peltier)-cooler mode   Thermogenerator mode

Figure 3.20

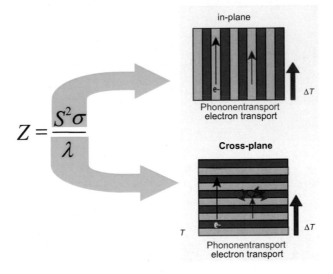

Figure 3.21

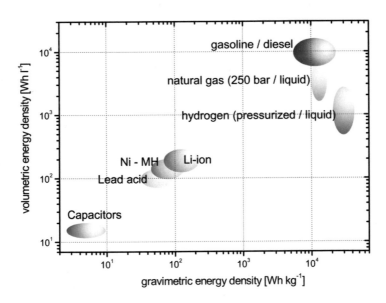

Figure 3.28

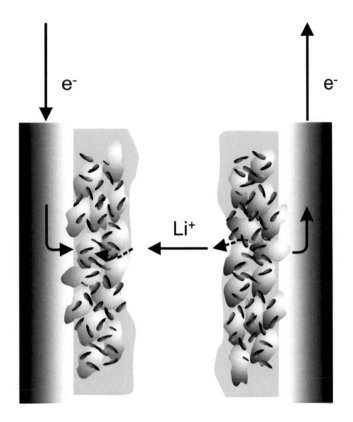

Figure 3.31

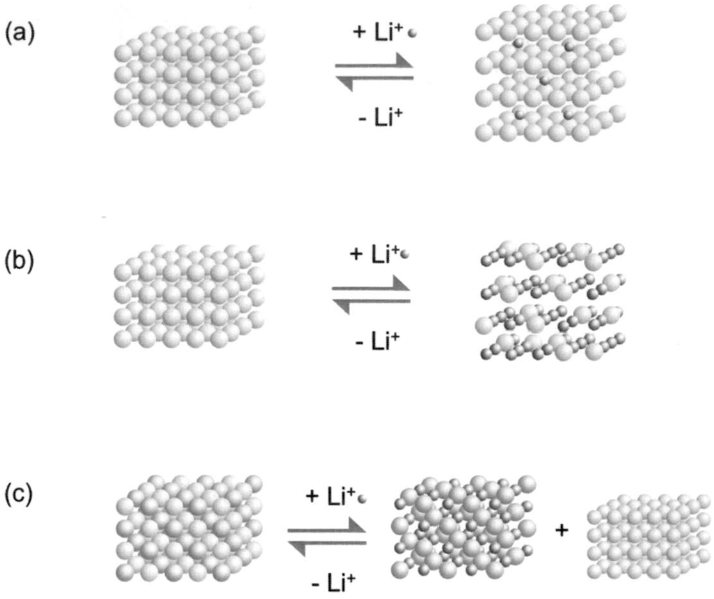

(a)

+ Li⁺

− Li⁺

(b)

+ Li⁺

− Li⁺

(c)

+ Li⁺

− Li⁺

+

Figure 3.32

Figure 3.33

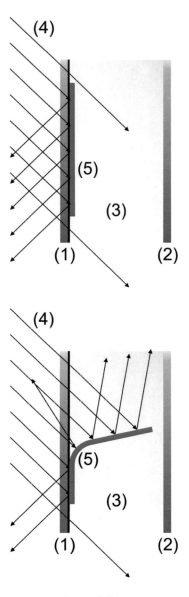

Figure 3.34

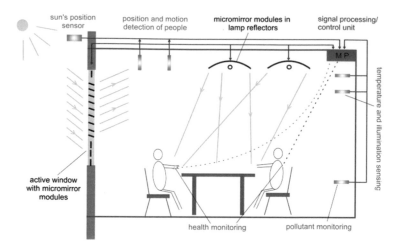

Figure 3.35

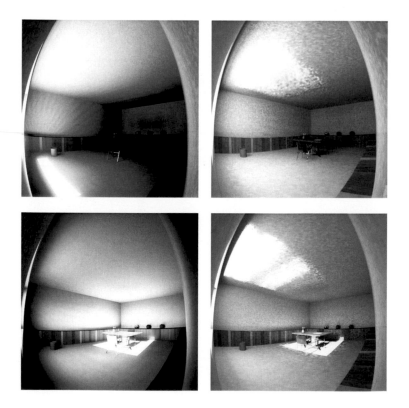

Figure 3.36

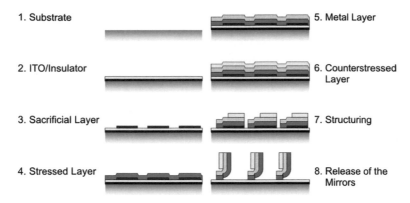

1. Substrate

2. ITO/Insulator

3. Sacrificial Layer

4. Stressed Layer

5. Metal Layer

6. Counterstressed Layer

7. Structuring

8. Release of the Mirrors

Figure 3.37

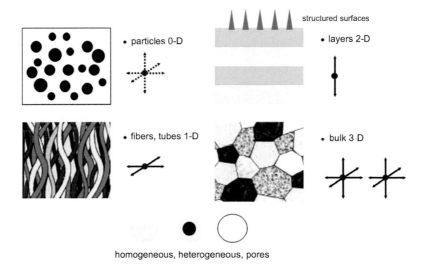

- particles 0-D

structured surfaces

- layers 2-D

- fibers, tubes 1-D

- bulk 3 D

homogeneous, heterogeneous, pores

Figure 3.38

Figure 3.40

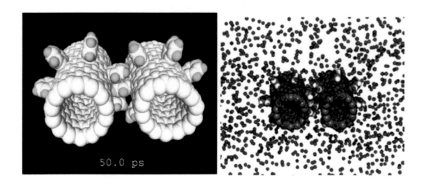

Figure 3.42

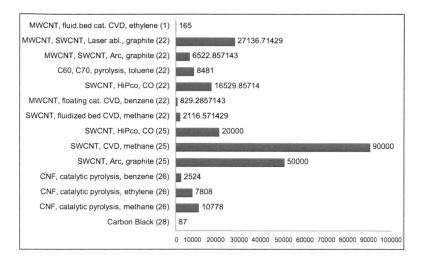

Figure 4.9

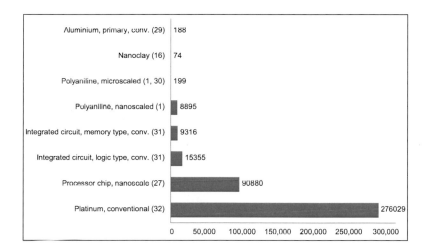

Figure 4.10

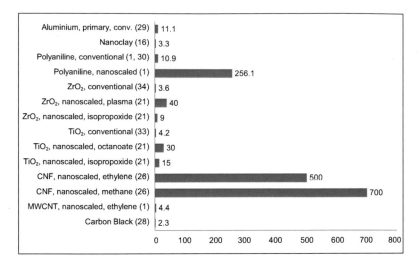

Figure 4.11